SYNTHESIS TECHNIQUES AND OPTIMIZATIONS FOR RECONFIGURABLE SYSTEMS

SYNTHESIS TECHNIQUES AND OPTIMIZATIONS FOR RECONFIGURABLE SYSTEMS

by

Ryan Kastner
Adam Kaplan
Majid Sarrafzadeh

KLUWER ACADEMIC PUBLISHERS
Boston / Dordrecht / New York / London

Distributors for North, Central and South America:
Kluwer Academic Publishers
101 Philip Drive
Assinippi Park
Norwell, Massachusetts 02061 USA
Telephone (781) 871-6600
Fax (781) 871-6528
E-Mail <kluwer@wkap.com>

Distributors for all other countries:
Kluwer Academic Publishers Group
Post Office Box 322
3300 AH Dordrecht, THE NETHERLANDS
Telephone 31 78 6576 000
Fax 31 78 6576 474
E-Mail <orderdept@wkap.nl>

    Electronic Services <http://www.wkap.nl>

Library of Congress Cataloging-in-Publication

CIP info or:

Title: Synthesis Techniques and Optimizations for Reconfigurable Systems
Author (s): Ryan Kastner, Adam Kaplan and Majid Sarrafzadeh

ISBN 978-1-4419-5414-5

Dedication

To my family – R. K.

To my wife Kim – A. K.

To Bita – M. S.

Contents

Preface

Reconfigurable systems are relatively new. The methods with which they are programmed and utilized have been discovered and refined all within the last decade. Though dreamed about decades ago, we have only recently begun to build and utilize these chameleon machines, whose logic can metamorphose within milliseconds. As computing elements, their potential performance is superior to that of software for many applications, yet they do not yield the higher performance of a completely fixed hardware approach. Thus reconfigurable devices represent a tradeoff between flexibility of software and the performance potential of traditional hardware. The utilization of reconfigurable logic and the means by which applications are mapped to this logic each require a rethinking and unification of the design methodologies used to create both software and hardware. In other words, reconfigurable devices are reconfiguring the way we approach system design.

The systems considered in this book could be either fully or partially built with reconfigurable technology. Often, it is useful to pair one or more reconfigurable device(s) with an ASIC or a microprocessor executing software (including meta-tasks such as reconfiguration of the devices). Additionally, it may be useful to embed small amounts of reconfigurable logic as core components within larger systems. In the face of such complex systems, it is infeasible to begin our design path at the logic level or even the hardware description level. Thus, a higher level of abstraction must be achieved.

Reconfigurable devices provide us with a great gift of flexibility, as their logic may change as needed to suit the changing needs of a system. It is imperative that we gather information about the entire application (or set of applications), and customize our system to its (their) needs. Application analysis begins with high-level code analysis, which allows us to exploit and extend over half a century of previous work in compilation of imperative programming languages. Our model of a reconfigurable system must begin here, at the compiler level, and be linked directly to models and methods used to synthesize both hardware and software applications.

This book discusses methods used to model reconfigurable applications at the system level, many of which could be incorporated directly into modern compilers. The book also discusses a framework for reconfigurable system synthesis which bridges the gap between application-level compiler analysis and high level device synthesis. The development of this framework (discussed in Chapter 5), and the creation of application analyses which further optimize its output (discussed in Chapters 7, 8, and 9), represent over four years of rigorous investigation within UCLA's Embedded and Reconfigurable Laboratory (ERLab) and UCSB's Extensible, Programmable and Reconfigurable Embedded SystemS (ExPRESS) Group. The research of these systems has not yet matured, and we continually strive to develop data and methods which will extend the collective understanding of reconfigurable system synthesis.

The material in this book assumes a basic understanding of logic design, hardware synthesis (from high-level architecture generation down to placement and routing), and the structure and form of high-level application constructs (such as loops and branches). However, this book may be read and used in the absence of such background knowledge. This text is aimed at researchers and system-level designers (both academic and industrial), but could easily be used as the text of a graduate-level course on reconfigurable system synthesis techniques.

It is our hope that the techniques applied within our studies, as well as our derived results, are both educational and useful to you, the reader of this book. Additionally, we hope that we have presented the subject matter in a clear and concise way. Comments, corrections, and suggestions are invited.

Chapter 1

INTRODUCTION

1.   OVERVIEW OF BOOK

The invention of the transistor in 1947 is arguably the most important discovery of the 20th century. Without it, Jack Kilby and Robert Noyce could not have created the integrated circuit (IC) in the late 50s. Within the span of 40 years, integrated circuits have permeated our world in almost every possible aspect. It is almost inconceivable to go through the day without the assistance of an IC. Computing devices play a part in many everyday tasks; they facilitate communication (e.g. cell phones, email) and allow us to organize our lives (e.g. digital assistants). To some extent, they even help keep us alive. Computing devices are used in a variety of different medical applications – from doing efficient DNA searches to keeping our heart beating (pace makers).

There are many factors that play a role in delivering computing devices to the world. Of course, no computing system could be made without an understanding of the underlying physics behind the transistors and other circuitry that provide the basic functionality of the IC. Another significant area for developing computing devices is the design and analysis of the organization of the computing device. The organization of computing systems is especially necessary as the complexity of the devices has exponentially grown since the development of the first IC. It is inconceivable to design a present-day computing device using the transistor as your basic element. Therefore, abstraction is needed to tame the complexity.

Abstraction is underlying view or model that a designer uses to create a computing device. As the size of computing systems has grown, so has the level of abstraction. Each step up in abstraction brings about the need to develop automated methods for synthesizing (the process of moving from one level of abstraction to a lower level of abstraction) the computing device. We have reached the point where synthesis techniques have matured to the point of the system level of abstraction. In order to move to successfully move to yet an even higher level of abstraction, we must enhance the synthesis techniques at the system level of abstraction. Our book focuses on this realm of abstraction, taking into account a special type of computing system – reconfigurable systems.

Reconfigurable systems initially emerged as fast prototyping device for ASIC designers. It allowed the designer to "compile" the application to reconfigurable hardware (e.g. FPGA) to determine that the application exhibited the correct functionality. The prototyping removed the costly step of fabricating the application, especially when fabrication yielded a device that exhibited incorrect functionality. Additionally, the rapid prototyping lessened the need for intense simulation to verify correctness. If the application functioned correctly in the environment when compiled to an FPGA, it would retain the same correctness once it was fabricated. The main drawback of the FPGA was the performance. An FPGA was far behind the ASIC in terms of important performance aspects, like latency, power consumption, etc. An application implemented on an FPGA was synthesized to the "static" nature of an ASIC and was not taking into account the dynamic reconfigurability allowed by the FPGA; in this sense, the performance of the FPGA can never overcome that of an ASIC.

The power of reconfigurable systems lays in the immense amount of flexibility that it provides. The flexibility allows run time reconfiguration and on-the-fly reorganization of the circuit based on the input parameters. Due to the ability to customize to the input data, many applications show speedups when implementing them on reconfigurable systems [1, 2]. In addition, many reconfigurable machines have achieved 100x speedups and 100x performance gain per unit silicon as compared to a similar microprocessor [3-7].

Reconfigurable devices have emerged as flexible, high-performance component in computing systems. We are seeing various examples of computing systems that are fully reconfigurable at the logic level as well as devices that are reconfigurable at the architectural level. In addition, we are seeing the increased use of reconfigurable cores as components in embedded systems – microprocessors coupled with a reconfigurable component as well as ASICs coupled with a reconfigurable component.

2. ORGANIZATION OF BOOK

This book focuses on methods needed for the synthesis and optimization at the system level of abstraction. In particular, we focus on synthesis techniques and optimizations for reconfigurable systems.

Chapter 2 gives an overview of the synthesis of digital systems. We focus on system synthesis and architectural synthesis as the remaining chapters lie between these two realms of abstraction.

Chapter 3 describes some concepts behind reconfigurable computing. We discuss the benefits of reconfigurable systems and survey a wide variety of reconfigurable computing architectures.

Chapter 4 talks about different formal models for specifying an application. We discuss some of the more important properties that must be modeled for embedded and reconfigurable systems. We briefly talk about some classic models of computation and present some of the more recent models of computation for embedded systems.

Chapter 5 presents a framework for system synthesis and describes different optimizations that can be done in order to improve the quality of the final mapping of the application to the reconfigurable devices. The remaining chapters discuss optimizations done within the framework to improve the quality of the final hardware implementation.

Chapter 6 looks at the hardware/software system partitioning problem. We start by describing the fundamental partitioning problem. Then, we discuss the hardware/software partitioning problem and survey several different algorithms for solving the problem.

Chapter 7 focuses on the problem of instruction generation. It relates the problem to several types of architectures – including microprocessors – but focuses on the use and impact of instruction generation to reconfigurable architectures.

Chapter 8 looks at the problem of data communication within a reconfigurable system. We utilize the compiler technique of static single assignment to minimize the amount of data that must be communicated between various components of our design. In addition, we give several methods to minimize the data communication and hence yield a design with smaller area.

Chapter 9 looks into the important problem of increasing the amount of parallelism in the design. It delves into two EPIC (Explicitly Parallel Instruction Computing) compiler techniques, superblock formation and trace scheduling, which are known to increase the amount of available parallelism in the application. Also, we discuss the side effects of these optimizations on other circuit parameters and especially focus on the effect of these techniques on the interconnect area. Furthermore, we give a formulation that combines hardware partitioning with the increase in parallelism.

Chapter 2

SYNTHESIS OF DIGITAL SYSTEMS

1. OVERVIEW

There are several fundamental levels of abstraction in hardware design. In the early days of hardware design, the entire circuit was mapped by hand. Soon, low-level standard cells (at the Boolean functional level of complexity) were used as basic building blocks. CAD algorithms for placement and routing are used to layout the circuit. As the number of transistors on a chip continued to increase, the level of abstraction moved to the logic level, microarchitectural level and the architectural level. The increasing complexity of computing systems brings the need for automation/exploration algorithms at an even higher level of abstraction – the system level.

In order to understand the stages of the synthesis of digital systems, we must understand the relationship between the different levels of abstraction. Gajski and Kuhn [8] introduced the Y-chart to describe such relationships. There are three different types of representations – behavioral, structural and physical. Each of these representations is a "branch" of the Y. The circles denote the level of abstraction. By transitioning from a circle (level) to one of its inner circles (levels), we perform a step called *refinement*. Refinement gives more details about the digital circuit. If we go the opposite direction, the process is called *abstraction*. Abstraction gives a higher view of the circuit, allowing more dramatic changes to occur. A *model of computation (MOC)* is a formal method of viewing and manipulating the circuit.

We can also move between the various representations. *Synthesis* is the transformation from a behavioral representation (at any level) to a structural representation of the same level. The reverse process – moving from a structural to behavioral representation – is called *analysis*. Going from a structural representation to a geometric representation is called *generation*. *Extraction* is the reverse of generation – creating a structural representation from a geometric representation.

It is possible to modify the representation while staying at the same level. Such modifications are called *optimizations*. There are many possible optimizations that can be done at the various levels and representations. Optimizations are done in order to improve some aspect of the circuit. For example, we may wish to optimize the circuit to reduce the power consumption, increase the throughput or minimize the area of the circuit. Optimizations are the heart of digital circuit synthesis. Determining the best area, power, and/or throughput for a circuit is a difficult problem. This is the problem of design space exploration.

The innermost circle is the physical level; this is the lowest level of abstraction. The geometric representation of a physical level is a set of polygons. These polygons correspond to metal wires, transistor contacts and channels, positive and negative doped regions of silicon and all other physical materials needed to realize transistors on a mask. Differential equations are a common way to describe the behavioral representation at the physical level. For example, the relationship between the inductance, voltage and capacitance of the circuit elements can be described in a behavioral manner using differential equations. As a structural representation at the physical level, we have transistors or we could even describe things using resistors and capacitors.

The next level of abstraction is the logic level. In this level, the behavior is often described using Boolean equations and Boolean logic. The typical basic elements for the structural representation are gates and flip-flops. The gates represent some (usually simple) Boolean function such as a 2-input "and" gate, a 4-input "or" gate, etc. The geometrical representation uses standard cells, which are usually modeled as rectangles. With each standard cell, there is an associated physical representation. For example, a 2-input "and" gate will have the transistor and/or polygonal representation.

By going one level higher in abstraction, we reach the microarchitecture level; this level is sometimes called register-transfer (RT) level or behavioral level. Some basic structural elements are data path (e.g. ALUs, adders, multipliers), memory elements (e.g. registers), and steering logic (e.g. multiplexors (MUXs)). An example behavioral representation at the microarchitecture level is a register transfer specification. Macro cells can be used to model the geometric representation.

The architecture level is the next level of abstraction. An algorithm is used as the behavioral representation. The algorithm allows common operators such as add, multiply, shift and control flow constructs like "for" and "while" loops and if-then-else branching. A processor is an example element as a structural representation at the architecture level. We use the term "instruction level" as being equivalent to the architecture level. "Instruction level" is a familiar term in computer architecture and assembly programming fields. We view a specific instruction set architecture (e.g. x86, alpha) an architectural level representation. The geometric representation for the architecture level is on the order of blocks or even chips.

The highest abstraction is the system level. The system level has many different models as the behavioral representation. We will go into more details those models in a following section. The structural representation has processing elements like a CPU; the processing elements are usually easily programmable. The system level geometric representation is on the level of chip or even a PCB board.

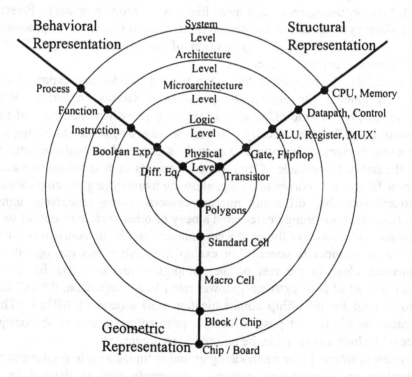

Figure 1 Gajski and Kuhn's Y-chart. It describes the relationship between the three representations of digital systems and the levels of description. The different points give an example representation at each level of abstraction.

In this chapter, we describe the fundamental ideas behind system synthesis. Additionally, we quickly discuss the basic concepts behind architectural synthesis. There is a direct interaction between system and architectural synthesis techniques. Therefore, it is important to understand the synthesis methods used at the architectural level.

2.   SYSTEM SYNTHESIS

Computing devices are permeating every portion of our world. From cell phones to laptops, embedded systems in planes, trains and automobiles to high-performance supercomputers, computing devices are everywhere. The rate of growth in the number of computing devices in the world continues to scale faster than the number of humans. We have begun to expect low cost, high-performance, specialized computing systems. Our cell phones must be light, have long battery, and good reception. Our personal digital assistants (PDAs) are becoming more and more like tiny desktop computers. Present day desktop systems and supercomputers run tens to thousands of processes at a time. With every new generation of computers, we expect better runtime, power, energy, size, etc., etc.

Fortunately, the number of transistors per die has doubled exponentially every 18 months for almost 40 years since Gordon Moore's initial observation [9]. A side effect of this amazing growth is that the cost per transistor has fallen dramatically. This allows cheaper, faster more complex computing devices. This amazing growth is expected to continue until at least the end of this decade. Computation elements such as microprocessors can now fit on a tiny corner of a chip, allowing memory (e.g. on chip cache), additional (possibly different) microprocessors, video processing units, reconfigurable computing devices and a bevy of other system resources on a single die. As a result of the on-die communication, the transmission of data between the devices is faster. For example, a system bus can operate at frequencies close to the rest of the computing devices (GHz frequency range), instead of at an extremely slower rate (As a comparison, the PCI bus protocol used for inter-chip communication runs around 66 MHz.). This increases the number of possibilities – equivalently the size of the design space – in which a system can be tailored to an application.

System synthesis is the methodologies and optimizations to implement an application as a computing system. An application is defined by a *specification*. A specification contains only the necessary requirement, constraints and functionality for the application. The system could be specified as a number of computational resources and the communication protocols between them or we may wish to completely derive the specification the system. The notion of system synthesis can encompass a broad area of system design. We could imagine a system that is comprised of various components – digital computing devices, analog devices, MEMS components, sensors, actuators, etc. The integration between these various components is often done in an ad hoc manner, most often using a system-by-system approach. We limit the scope of this book to systems consisting solely of digital devices.

A *platform* [10] is a physical realization of a system. The fundamental goal of system synthesis is the transformation the application (as defined by a specification) onto some given platform. Different platforms lie on different abstraction levels. In fact, a platform may lie on multiple levels of abstraction, e.g. an FPGA+CPU platform has a component on the logic and architectural levels of abstraction. Different platforms have different performance characteristics. Some platforms target high throughput or clock frequencies, while others are optimized for the lowest power consumption.

User demands for a variety of high performance, lightweight and low power computing devices have created a market for application specific computing systems. However, the design of a custom computing system takes considerable amount of time and a tremendous amount of resources. A well designed platform can provides a middle ground between time to market and system performance.

It is conceivable for a platform to contain a number of different programmable computing devices. For example, such a system level platform may consist of a RISC processor, VLIW processor and an FPGA (see Figure 1). Each of the devices in the platform allows different design tradeoffs when it is programmed. The maximum amount of parallelism of an FPGA is much larger than that of the VLIW processor, which in turn is larger than the RISC processor. On the other hand, an FPGA takes much longer to program as opposed to the processor counterparts.

A platform may be designed for a specific set of applications. For example, an audio platform may be designed for applications that predominately process audio, e.g. an MP3 player. This platform would undoubtedly have dedicated (ASIC) analog to digital and digital to analog converters. Additionally, it would have a variety of general-purpose processing devices like a DSP processor and reconfigurable device. The general-purpose elements add flexibility to the platform. Flexibility allows the platform to be used towards a variety of different applications. The aforementioned audio platform could be used in a MP3 player, cell phone, video game console, etc., depending on the amount of flexibility in the platform.

As a platform becomes more flexible, it can handle a larger number of applications making it more financially viable to fabricate. The non-recurring engineering (NRE) costs of fabrications grow substantially with every process change. The NRE costs alone keep many companies out of designing and fabricating application specific computing systems as a system on a chip (SOC). Yet, the only reason they would wish to fabricate an SOC is if some current platform does not suffice to their needs for performance, power, weight, etc.

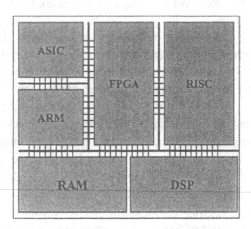

Figure 2 An example of a platform based computing system. There are 6 components to this platform – three microprocessors (RISC, ARM and DSP), an ASIC and a RAM. This platform uses a simple method of communication through a system bus.

Currently the design of system level platforms like SOCs is done manually in an ad hoc manner. We must look into efficient CAD methods for the design of platforms. Furthermore, we must develop automated methods for mapping a specific application to a platform. A translator from the application specification to hardware description is a necessary component for both the design of a platform and the mapping of an application to a platform.

2.1   Architectural Synthesis

Architecture synthesis is the process of creating a structural microarchitectural level representation from a behavioral architectural description of an application. A *structural representation* defines an exact interconnection between a set of architectural resources. An *architectural resource* is a memory, data path, or steering logic. A memory component is a method of storing the state of the circuit. A register is an example of memory component. A data path component performs an arithmetic or logic operation e.g. an ALU, multiplier, shifter, etc. Steering logic is used to route data. For example, a multiplexor propagates a particular piece of data (correspondingly a set of signals) depending on a condition. A control unit (controller) issues control signals to the direct the resources.

Architectural synthesis can be performed using any number of different methods. Additionally, a designer can add additional constraints or optimization objectives. For example, we may want to produce a circuit that requires the least amount of area. In this case, our objective function would be to minimize the area. Many other constraints have been considered during architectural synthesis. Throughput, power, clock frequency, and latency are some of the more popular optimization variables.

The architectural synthesis problem can be defined in the following manner: given a CDFG, a set of fully characterized architectural resources and a set of constraints and an optimization function, determine a fully connected set of resources (a structural representation) conforms to the given constraints and minimizes the objective function. The architectural synthesis problem can be split into two sub-problems: scheduling and resource binding.

Scheduling determines the temporal ordering of the vertices in the CDFG. Given a set of operations with execution delays and a partial ordering, the scheduling problem determines the start time for each operation. The start times must follow the precedence constraints as specified in the CDFG. Additional restrictions such as timing and area constraints may be added to the problem, depending on the architecture one is targeting.

Two important scheduling algorithms are the *as soon as possible (ASAP)* and the *as late as possible (ALAP)* algorithms. The ASAP algorithm has run time $O(|V||E|)$. The vertices are topologically sorted. This insures that when a vertex is visited, all of its predecessors have been visited. The start time of a vertex earliest possible time the operation can be started without violating the precedent constraints. The ASAP algorithm optimally solves the unconstrained scheduling problem. Just as the ASAP algorithm gives the earliest possible time that an operation can be scheduled, the ALAP algorithm gives the latest possible time that an operation can be scheduled. Given a maximum latency, the ALAP algorithm starts by performing a reverse topological sort of the vertices of the CDFG. The start time of a vertex is determined by finding the minimum start time of all of the operations successors.

The *slack* of an operation is the difference in between the start time of the operation scheduled by the as late as possible (ALAP) algorithm and the as soon as possible (ASAP) algorithm. The slack gives an idea as to how much freedom the operation has to be scheduled.

The scheduling problems becomes NP-Complete [11] when resource constraints are added. There are many different kinds of constraints that we can put on the scheduling problem. We could schedule for minimum latency given a set of resource constraints. Or we could consider the dual of that problem, which attempts to minimize the number of resources under latency constraints. When we assume that a single type of resource can be used for any operation, the problem becomes a precedence-constrained multiprocessor scheduling problem. This problem is intractable and remains intractable even when all the resources have unit delay.

Hu [12] provided a lower bound on the number of resources for the precedence-constrained multiprocessor scheduling problem. Hu also describes a greedy algorithm that gives an optimal result when the CDFG is a tree. It starts at the first time step and schedule the vertices on the longest path first. It continues until all of the operations are scheduled.

Unfortunately, the CDFG is not often a tree and no exact polynomial time algorithm exists to solve the general scheduling problem. As a result, many heuristics have been proposed to solve the problem.

One popular scheduling heuristic is list scheduling. The list scheduling algorithms use a priority queue to hold the unscheduled nodes. At each step, an operation is popped from the queue and scheduled. This continues until all of the operations have been scheduled. The priority queue is sorted based on some measure of urgency to schedule the operation. A common sorting function is based on sorting the vertices corresponding to the distance to the sink. The list scheduling algorithm runs in linear time, making it a quite attractive method of scheduling.

Force directed scheduling is another class of scheduling heuristic. It was first proposed by Paulin and Knight [13] in 1989. The force directed algorithm works in a constructive manner. They proposed two algorithms for force directed scheduling. One algorithm minimizes latency under resource constraints and the other minimizes resources under latency constraints. The algorithms can be differentiated by the order in which they choose to schedule the operations. First consider scheduling under latency constraints. The algorithm schedules all of the available resources at each time step during an iteration. At each time step, the resources are matched with operations that have the largest force. The minimum resource algorithm schedules the operations based on the calculation of the force on the operations. At each step, it takes the operation with the lowest force and schedules it. Essentially, it tries to schedule operations to minimize the local concurrency.

Other scheduling algorithms have been proposed that employ an iterative improvement technique. The technique starts with an initial scheduling and transforms the schedule by changing the start time of some of the operations. Trace scheduling [14] and percolation scheduling [15] are two examples of iterative improvement scheduling algorithms.

Resource binding is the assignment of hardware resources to one or more operations; it is an explicit mapping between operations and resources. The goal of resource binding is to minimize the area by allowing multiple operations to share a common resource. Resource binding may be done before or after scheduling. It is most often done after scheduling, so we will focus on that methodology. When it is done after scheduling, the scheduling will give some limitations on the possible resource bindings. For example, operations that are scheduled to execute during the same time step cannot share the same resource. To be more precise, any two operations can be bound to the same resource if they are not concurrent, i.e. are not scheduled in overlapping time steps, and the operations can be implemented using the same resource. For example, if a resource is an ALU, both an addition and subtraction operation can be bound to an ALU resource.

The resource binding can greatly affect the area and latency of the circuit as it dictates the amount of steering logic and memory components of the circuit. We first consider the slightly simplified problem of resource binding for circuits that are dominated by the arithmetic units.

When the circuit is arithmetically bounded, the delay and area of steering and memory components are miniscule when compared to the arithmetic units. One method of modeling the problem is through a resource compatibility graph. Two operations are *compatible* if they can be scheduled using the same resource. The *resource compatibility graph* has a one-to-one correspondence between the operations and the resources of the graph. There exists an edge between two vertices if the corresponding resources are compatible.

Since we are focusing on the arithmetically bounded problem, an optimum resource binding is one that minimizes the number of resources needed to bind the operations. This is equivalent to finding the minimum number of cliques in the resource compatibility graph. A conflict graph is the complement of a compatibility graph. Minimizing the number resources can be cast as a minimum clique partitioning (each partition must be a clique) problem of the compatibility graph or into a minimum coloring problem of the conflict graph [16]. The resources can be functional units (resource allocation), registers (register allocation), etc. The graph coloring problem for the conflict graph of an interval graph can be solved in polynomial time via the left_edge algorithm [17].

3. SUMMARY

This chapter discussed the basic terms related to the synthesis of digital systems. We briefly described different abstractions and representations of digital systems. We focused our discussion on synthesis methods at the system and architectural levels of abstraction. We discussed algorithms for solving two fundamental problems – scheduling and resource allocation.

We defined some basic terminology and provide basic information about architectural synthesis. Our coverage of that material is minimal.

Furthermore, we do not discuss logic or physical synthesis techniques that are essential to fully realize the circuit. These topics are out of the scope of this book. We refer the interested reader to the following books [16, 18, 19] for more information on these topics. The remainder of this book focuses on the system level synthesis techniques for reconfigurable systems.

Chapter 3

RECONFIGURABLE SYSTEMS

The terms "reconfigurable computing" and "reconfigurable systems" have been used somewhat judiciously ever since the introduction of the field programmable gate array (FPGA) in the mid-eighties. The term has been used to describe logic level devices such as FPGAs, complex programmable logic devices (CPLDs), as well as devices that look more like arrays of processors. In this chapter, we attempt to present a clear definition of the term "reconfigurable". We start by explaining some of the basic concepts behind reconfigurable systems. Then, we examine some of the immense benefits that reconfigurable systems have shown over a wide range of applications. Finally, we survey a variety of different reconfigurable architectures. We give a methodology to classify the systems according their computational level of abstraction.

Note: Several terms will be used in this section without definition, e.g. FPGA, PLD, CPLD, PLA, PAL. For a precise definition, see Section 2.

1. WHAT IS RECONFIGURABLE?

It is impossible to talk about reconfigurable systems without first explaining what it means to "reconfigure" or how to go about the process of reconfiguration. The term "reconfigurable" is rather elusive; it has been used to describe many different things. The following are just a few of the definitions for reconfigurable computing:

"Reconfigurable computing, in the abstract sense, refers to any information-processing system in which blocks of hardware can be reorganized or repurposed to adapt to changing data flows or algorithms"
 - Ron Wilson, EE Times

"Reconfigurable computers are those machines that use the reconfigurable aspects of reconfigurable processing units and field programmable gate arrays to implement an algorithm. The algorithms are partitioned into a sequence of hardware implementable objects. These hardware objects represent the serial behavior of the algorithm and can be executed sequentially. The use of hardware objects gives the developer a logic-on-demand capability that is the basis of configurable computing."
 – Virtual Computer Corporation (VCC)

"A reconfigurable computer is a device which computes by using post-fabrication spatial connections of compute elements."
 – Andre DeHon

"Reconfigurable devices contain an array of computational elements whose functionality is determined through multiple programmable configuration bits."
 – Katherine Compton and Scott Hauck

"General purpose custom hardware"
 – Seth Copen Goldstein

"'On the fly' ASIC"
 – Fadi Kurdahi

We can attempt to find the "true" meaning of reconfigurable computing by consulting a dictionary. Of course, a standard dictionary does not list the term "reconfigurable computing"; in fact, the Merriam-Webster dictionary does not even have an entry for the "reconfigurable". However, it does have an entry for the prefix "re" and the root word "configure"

Main Entry: **re-**
Function: *prefix*
1: again : anew <*retell*>
2: back : backward <*recall*>

Main Entry: **con·fig·ure**
Function: *transitive verb*
: to set up for operation especially in a particular way

From this, we can ascertain that "reconfigure" means "to set up for operation again", which is a good start towards a more precise definition of reconfigurable computing.

We believe that it is hard to find an essential definition of reconfigurable computing. An essential definition requires that we give the properties shared by everything that capable of reconfiguration. Instead, we aim to give an ostensive definition – a definition constructed by examples. By understanding some simpler definitions relating to reconfigurable computing, we can gain more insight into the realm of reconfigurable devices.

1.1 Basic Underlying Concepts

This section will describe some of the basic terminology behind reconfigurable systems. In particular, we explain the concept of configurability and reconfigurability. Additionally, we discuss the relationship between reconfigurability and programmability. Finally, we look at the tradeoff between temporal and spatial computation.

1.1.1 Configure

What does it mean when we configure a device? Configurability is quite clear when we are referring to a programmable logic device (PLD); a PLD is configured once the device is programmed using a bitstream. By programming the PLD, we are setting the functionality of the device. In the same sense, a microprocessor can be programmed by setting it to perform an instruction from the instruction set architecture. Following this line of intuition, we say that a device is *configured* when its functionality is set. This definition is related to DeHon's definition of the binding time of a device. He defines *binding time* as the time when the design is configured or customized.

One interesting side effect of using this definition is that one could view the process of designing an ASIC as configuring the ASIC. The functionality of the ASIC is set when the circuit is fabricated; in other words, ASICs bind function to silicon at fabrication time. Using the notion of levels of abstraction, we can view ASICs as being configured at the physical level. We will soon see that configurability at such a low level of abstraction has advantages and disadvantages.

Configurability also lies at other levels of abstraction. For example, gate arrays are configured when the routing is defined, e.g. when the fuse is blown creating a connection between two wires. The functionality of the gate array is set once all the necessary fuses are blown. We can view this as configurability at the logic level. A microprocessor is configured when an instruction is read from memory. Therefore, microprocessors bind functionality at every cycle. The functionality is defined by the currently executing instruction, which is pulled from restricted set of instructions – the instruction set architecture (ISA). The configurability of a microprocessor is defined at the microarchitectural level of abstraction. There are many possible binding times for a computing device. A device can be bound at design time, during the deployment, between executing phases and during execution.

1.1.2   Reconfigure

Now that we have defined configuration, we proceed to the definition of reconfiguration. Referring once again to the Merriam-Webster dictionary, the prefix "re" can take the meaning "again" or "anew". Therefore, reconfigure means to configure again or to configure anew. We define *reconfigurable* as the ability to continually change the functionality of the device.

Programmable logic devices (PLDs) like FPGAs and CPLDs are reconfigurable, as they can be reprogrammed to change their functionality. These devices are configurable at the logic level of abstraction. Mask or via programmable devices, e.g. PAL, PLAs, are configurable, but not reconfigurable; the functionality of these devices can be set exactly once. This happens when the fuse is blown in a PAL, PLA and when the routing layers are decided as in mask programmable devices.

Our definition implies that microprocessors are reconfigurable devices. However, a microprocessor is not normally considered a reconfigurable device. Before we make a distinction between microprocessors and reconfigurable devices, let's look at the relationship between programmability and reconfigurability.

1.1.3   Programmability

Programmability is the capability of computing any of a number of fixed functions after fabrication. A fixed functional unit computes exactly one function. For example, a floating point multiplier and a Booth adder are both fixed functional units. They only have the ability to compute multiplication and addition, respectively. In contrast, an arithmetic logic unit (ALU) has the ability to compute a number of functions. It can compute an addition, subtraction or a comparison operation. All of these operations are done using the same circuitry. The exact operation that ALU computes depends on the way that it is programmed. The basic functionality of the

ALU is addition. However, it can be set to do a subtraction by inverting the bits of the second operand and setting the carry in bit. A comparison operation is done by looking at the result of subtracting the two operands. The exact function of the ALU is set after fabrication. The ALU is used in almost every instruction in a microprocessor ISA. It is used in arithmetic instructions as previously described and it is used in memory instructions to calculate the memory address that is accessed.

The "amount" of programmability is related to the number of different functions that a device can implement. An ASIC can perform exactly one task – the task it was fabricated to perform. A general purpose processor (GPP) can perform a wide variety of operations. The instruction set defines the number of different operations that the processor can perform at any given time. The functionality of a PLD is dictated by the bitstream used to program it. A PLD has extremely fine grained, logic level control. For every gate equivalent of a PLD, there are on the order of 10 to 100 programming bits. The size of a bitstream of the current high performance PLDs is approximately 43 million (Xilinx Virtex-II Pro XC2VP125 has 43,602,784 configuration bits) compared to the 64 bits needed to program a current top of the line GPP.

A programmable device has many benefits over ASICs and other non-programmable devices. A programmable device allows application development after fabrication. The functionality of a programmable device can vary widely across many different application domains. Additionally, a programmable device allows for non-permanent customization. The functionality of the device can change over time. This is especially important for applications with rapidly changing standards (e.g. wireless protocols). Programmability allows for prototyping the application during development. We can implement different revisions of a design on programmable device until we get a field-tested, correctly functioning device.

The economies of scale favor programmable devices. As fabrication becomes more and more expensive, only devices with a high amount of volume are profitable. The International Technology Roadmap for Semiconductors predicts that designs with 50 nm feature size will be feasible by 2007. However, it predicts that the cost of the masks and exposure systems needed to fabricate these chips will likely impede the development of these chips. Current day designs (below 150 nm feature size) already cost over one million dollars to fabricate.

Programmable devices can be used to implement a wide variety of applications, whereas ASICs implement precisely one application. Since a programmable device can implement a wide variety number of applications, the large cost of fabrication can be amortized over a large number of users. This decreases the price per device for low volume applications, making programmable devices the most attractive option for low volume applications. However, once the volume of the application reaches a certain point (called the *crossover point*), an ASIC becomes a more attractive option for implementing the application. At that point, it is better to pay for the fabrication cost and design the chip yourself. Referring to Figure 3, the cost per unit of programmable devices is larger than that of ASICs. Correspondingly, the slope of the programmable device line is larger than the ASIC line. The increased unit cost of a programmable device is the price charged by the semiconductor company that manufactures the device.

The crossover point is the breakeven point between manufacturing the device and purchasing a manufactured device. After that point, it is financially beneficial for the application designer to pay the NRE costs in order to get the cheaper per unit cost, i.e. eliminate the middle man. Increasing manufacturing costs increase the NRE costs, making the crossover point increase with every new technology node. Therefore, programmable devices are an increasingly attractive option for future application designers. Additionally, it allows low volume applications to use the latest manufacturing techniques.

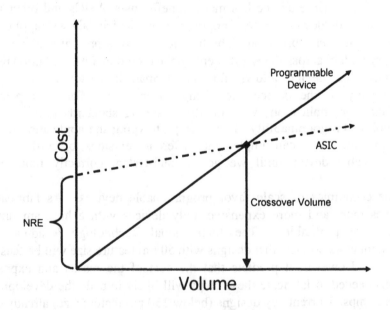

Figure 3 Cost versus volume comparison of programmable devices and ASICs.

Obviously, such a simple model has its drawbacks. It does not take into account other metrics of interest for the application designer. For example, high performance applications often need the best performance in terms of delay, throughput, energy, etc. Additionally, it is often hard to quantify NRE costs and the price per unit. Often, the true cost (in terms of benefit to the user) of the device is a complex function of many different variables. Additionally, both applications and devices often have hard and soft constraint, For example, the application must be battery powered, the device only obeys certain protocols, etc. This makes it quite difficult to determine the best device for the application. Despite these limitations, the simple model shows a clear trend for the further emergence of programmable devices for future applications.

The time to market of a computing device is another large factor in the profitability of the device. A device that reaches the market quickly (before the devices of the competitor) will invariably have a larger demand and make a larger profit than a device that hits the market months later (even if that device is better). Programmable devices are already fabricated, hence they do not require the additional, lengthy fabrication time. The lack of fabrication time could be the difference between a successful (profitable) and unsuccessful device.

Many of the properties a device are related to its programmability. The flexibility of the device is positively correlated with the number of bits used to program the device. The more programmable the device, the more flexible that the device becomes. As an example, let us compare and FPGA to a microprocessor. The FPGA has orders of magnitude more programming bits than a microprocessor. Additionally, an FPGA can implement any function performed on a microprocessor (consider the fact that you can implement a soft core microprocessor on an FPGA), whereas the opposite is not true. These series of observations point towards a positive correlation between the number of programming bits and flexibility. Though undoubtedly counterexamples can be found, we believe this general trend holds across different classes of programmable devices. Programmability is also related to the performance of the system. It is estimated that custom logic is one order of magnitude faster and up to two orders of magnitude smaller than an equivalent PLD solution [20]. Also, programmable devices introduce increased power dissipation and energy usage compared with an ASIC.

In summary, programmable systems have many benefits over non-programmable systems (e.g. ASICs). Programmability provides non-permanent customization and application development after fabrication. Economies of scale favor programmable systems as they allow the increasing, large, fixed design costs to be amortized over many people. Furthermore, programmable devices are already fabricated allowing quick time-to-market. Programmable devices can evolve to meet any emerging requirements, standards and new ideas that may occur over the lifetime of the device. On the negative side, programmability is associated with an increase in power and delay of the device. However, designers can mitigate these overheads through good design techniques.

1.1.4   Programmable vs. Reconfigurable

So what is the relationship between programmability and reconfigurability? The two terms share many similarities to the point where it is hard to make exact distinction between them. Reconfigurable devices were initially seen as a more flexible or programmable hardware alternative e.g. PLDs, whereas microprocessors are traditionally classified as programmable devices. A discussion of PLDs and microprocessors provide a good starting point for a comparison between programmable and reconfigurable devices. While there are some differences, we will show that these two terms share more similarities than differences.

A microprocessor shares many characteristics with a reconfigurable architecture. The processor architecture is general purpose; it can be programmed to solve any current or future task (programmed by some language like C/C++, Java, Fortran, etc.). Likewise, a PLD can be configured to run almost any application. One major difference between the two devices is that the processor architecture, unlike a PLD, is "spatially" fixed at the architecture level.

DeHon [21] describes a microprocessor as a device that makes connections in time. This is related to the fact that microprocessors have largely evolved as a general purpose computing device. They are almost solely programmed using sequential languages such as C/C++ Java and Fortran. These "software" programming languages are based on a load/store paradigm of computing where operations are done sequentially on various memory locations. The memory locations are used for communication between the various operations. The basic operations done by microprocessors are primarily limited input/output arithmetic operations with no state.

Microprocessors are optimized to the general case. This allows them to be used in a wide variety of applications. They are extremely flexible computing devices. However, microprocessors are often overly general for the applications that they execute. This over-specification comes in the form of overhead circuitry in controllers, unnecessary multiplexors and registers, which cause a significant degradation of overall performance. Often, the general purpose computing resources are poorly matched to the application. They may be too large (only need 8 bits instead of 32 bits). Extremely parallel applications are a poor match for execution on microprocessors due to the microprocessors sequential nature of execution.

a)

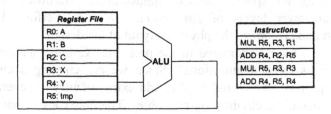

b)

Figure 4 Different model of execution for the expression $y = x^2 + Bx + C$. Part a) is temporal execution and part b) is spatial execution.

Spatial execution allows for operations to be performed in parallel. The parallelism is only constrained by the dependencies among the operations. Consider the arithmetic expression $y = x^2 + Bx + C$ (see Figure 4). This expression requires 2 multiplication and 2 addition operations. The microprocessor performs the operations sequentially using registers to hold the intermediate values. Using spatial execution, we can perform two of those operations in parallel. Furthermore, note that here, the connections are made through wires and not registers; connections are made spatially, not temporally.

Spatial execution is often equated with the hardware implementation of an application. When a design is implemented in hardware, we have the freedom to explore a range of different microarchitectures. We can increase or decrease the number of registers. We have the ability to specify different bitwidths for different operations. We can choose between the numbers and types of functional units that are needed. We can even design customized functional units for special purpose applications. Hardware is further optimized at lower levels of abstraction. We can perform bit level optimization and customize the physical layout of the design.

Hardware is fully customized to the problem. Communication can be implemented as direct connections. Since we are creating a circuit for exactly one purpose, we do not need to add extra circuitry to generalize the problem i.e. make the circuit programmable. The drawback of hardware is that it is inflexible. The functionality can not be changed after the circuit is fabricated.

PLDs allow spatial execution, while providing post-fabrication configuration. They combine some of the key characteristics of both software (microprocessors) and hardware (ASICs). They are fast, since they allow spatial parallelism and application specific operators and control circuitry. They are flexible since operators and interconnect are programmable.

The flexibility of the PLD comes at a cost. The programmability adds area and delay overhead in the form of switches, configuration bits and large programmable computation elements. Compiling an application description to a PLD is difficult because the programmability lies at such a low level of abstraction. Most PLD programming tools use techniques similar to hardware synthesis making compilation to PLDs extremely time consuming.

DeHon [22] provides an excellent comparison between microprocessors and PLDs using the notions of *instruction depth* and *datapath width*. Datapath width is defined as the number of compute operators at the bit level that are controlled by a single configuration. The instruction depth is the number of configurations that can be stored locally and changed on each operating cycle. A microprocessor has instruction depth on the order of megabytes (on-chip cache). It has a datapath width of 32 or 64 bits. On the other hand, the PLD has a datapath width around 1 (it works on the logic level) and can store only a small number of configurations on chip (typically only one at a time).

The performance of the microprocessor and PLD greatly depend on the application at hand. An application with 32 bit data and regular, sequential computation patterns will execute extremely efficiently using a microprocessor. On the other hand, applications with course grained parallelism, customized computations and bit level operations are better suited towards a spatial implementation on a PLD.

Reconfigurable computing is often associated with programmable devices that lie somewhere in between microprocessors and PLDs. This involves a large number of varying devices. In the next section, we look at different methods to further classify the reconfigurable devices. We focus on three particular characteristics of reconfigurable devices – granularity, coupling and dimensionality.

1.2 Characteristics of Reconfigurable Computers

1.2.1 Granularity

The *granularity* of a reconfigurable device is the abstraction level used to program or configure the device. We define the granularity using levels of abstraction. The granularity of a reconfigurable device may use a Boolean, instruction, function or process level representation.

The granularity is proportional to the length of a configuration. A low granularity means that the configuration is very long; a high level of granularity generally denotes shorter configurations. The length of a configuration is related to datapath width – the larger the data path width of a device, the larger the configuration needed to program the device. Also, the granularity gives some notion as to the underlying freedom of the device. Instruction level granularity (e.g. ISA) only allows you to specify a limited number of prespecified registers locations and small number of operations on those locations. Lower granularity level allows you to specify more locations and different, complex customized functional units.

FPGAs are examples of fine grain architectures. The primary computational element in FPGAs is a limited input lookup table. A lookup table usually has four inputs, meaning that you can use them to implement any four input function. This Boolean level granularity can be used to create more complex functions. We can implement circuits with irregular data width, e.g. an 18 bit multiplier or 24 bit adder. However, the Boolean level granularity is too fine to efficiently execute operations requiring a large number of inputs.

A coarser grain reconfigurable device consists of computational units that perform instruction level operations. The programmable computational units vary from byte-width (8 inputs) to word-width (32 inputs) datapath operations. The computational units rarely have state. They are usually read from registers and write from registers. The computational units often are optimized towards large computations like addition or multiplication. These computational units perform instruction level operations extremely efficiently. However, this level of granularity is quite inefficient at implementing Boolean functions or other limited input functions.

In general, the more closely the application data is matched to the granularity, the more efficient the device will execute the application. For example, a DSP application requires a lot of word-size addition and multiplications. Therefore, a device with instruction level granularity would provide the best performance. If an application requires a bunch of Boolean operations, then device with Boolean level reconfigurability would perform the task most efficiently.

Function level granularity is an even coarser than instruction level. Function level computational units are complex multicycle operations. The behavior consists of a number of different instructions with a large amount of control and data dependencies. Function level computation units consist of a number of different complex behaviors. The computational units contain state using registers and configurable interconnect; they can often be used in a pipelined manner.

The highest coarsest granularity that we consider is the process level reconfigurable devices. These devices implement are extremely complex and often take hundreds or thousands of cycles to complete. The computational units at the process level are sometimes very small microprocessors with local memory banks for storing data and instructions. These devices require applications that are extremely parallel in nature. Some of the process level reconfigurable devices are literally multiprocessor systems on a single die.

Often, applications have requirements for different types of parallelism. Many hybrid or heterogeneous reconfigurable devices have emerged, which combine reconfigurability at multiple levels of granularity. A recent trend in the design of high performance programmable logic devices is to incorporate components at higher levels of granularity in addition to providing Boolean reconfigurability. These devices have fixed point multipliers, embedded RAM, microprocessors and look up tables. We have seen many other successful hybrid reconfigurable devices, which we further discuss later in the chapter.

1.2.2   Coupling

The *coupling* of a reconfigurable device is the method used to integrate the device into the overall computing system. Our definition of coupling largely follows that of Compton and Hauck [23], though we adapt it some to conform more towards our terminology.

We can describe the coupling of the reconfigurable device using levels of abstraction. The loosest coupling comes at the system level. Here the reconfigurable devices act as an externally connected processing unit. This type of coupling typically has slow transfer rates between the devices; hence communication between the devices must occur infrequently. Ideally, the applications running on this system will communicate small infrequent amounts of data and the reconfigurable devices will perform a large amount of processing on that data. The large amount of processing permits longer device reconfiguration time, as the reconfiguration time is small compared to the overall processing time. There are many different architectures that have explored system level coupling of FPGAs and GPPs [5, 24].

Coupling can occur at the architectural level. In this case, the reconfigurable device is meant to act more like coprocessor. At the architectural level, the reconfigurable device communicates through memory or special register banks. The data transfer is much more local in this case. Here, we can afford to communicate more data compared to system level coupling. An architecturally coupled reconfigurable device often runs for hundreds of cycles. Some controller (perhaps a GPP) assigns function level behaviors to be implemented on the reconfigurable device. It also permits separate threads of control (task level parallelism).

The tightest coupling occurs at the microarchitectural level. Here, the reconfigurable devices are used as functional unit within a data path. This coupling has extremely fast communication, as registers or direct wire connections are used to relay data into the device. The reconfigurable device is used often, e.g. every cycle, so the configuration time of the device must be small. Coupling at this level often looks to use the reconfigurable device to exploit instruction or bit level parallelism.

A number of systems have explored coupling reconfigurable devices with a general purpose processor (GPP) [25-28]. Some of these interface through special registers or use special memory locations. Others are more tightly integrated at the microarchitectural level. They may share the register bank with the GPP.

The coupling partially determines the reconfiguration time of the device. A device that is tightly coupled (e.g. shares registers with the GPP) can not tolerate long reconfiguration times. In general, tightly coupled reconfigurable devices require short reconfiguration times in order to be efficient. Loosely coupled reconfigurable devices interface through slower mechanisms (consider the 66 MHz operating frequency of a PCI bus versus the GHz operating frequency of a register) and perform a large amount of computation. Therefore, they can tolerate longer reconfiguration times.

1.2.3 Dimensionality

The *dimensionality* of a reconfigurable device is the manner in which the processing elements communicate and are physically arranged. A reconfigurable architecture can be arranged in a variety of ways. Some use linear (one dimensional) communication. Many reconfigurable architectures are arranged in two dimensional arrays, while others are arranged in a mesh.

This section attempts to provide a classification scheme for reconfigurable devices based on dimensionality. We describe several different architecture arrangements utilized by reconfigurable devices.

Mesh based architectures arrange their processing elements in a 2-D array. The interconnect is usually restricted to a single dimension (either horizontal or vertical wire). The processing elements are connected in a nearest neighbor fashion i.e. there is rich local interconnection complexity with very few long interconnects.

Island style architectures use a two dimensional structure for both communication and layout of the processing elements. These architectures are similar to mesh based architectures. They differ in the way the processing elements communicate. Mesh based architectures have nearest neighbor communications schemes, while island style architectures allow global communication. An FPGA is an example of an island style architecture. The processing elements are arranged in a 2-D manner. Communications is done both locally and globally using routing segments of varying lengths. Processing elements that are far away can quickly communicate using the long segments, while local communication is done through the short routing segments.

Row based architectures have a 2-D layout of processing elements, though only allow for 1-D communication. The processing elements in a row are combined to perform a single operation. The rows in these architectures are atomically configured. For example, Garp [29] and Chimaera [28] use this type of architecture. Here, the processing elements are bit operations and each row is configured to do word level computations.

Some reconfigurable devices are arranged in a similar manner to systolic arrays. A *systolic array* processes data from memory in a rhythmic fashion, passing through many processing elements before returning the data to memory. Kung and Leiserson [30] pioneered systolic arrays in the late 70s as an architecture for VLSI circuits. They wanted to try to reduce possible complications in designing and manufacturing VLSI circuits.

A systolic array is a set of simple processing elements with regular and local connections. The array processing input data in a predetermined, pipelined manner. Simple processing elements are used to simplify the creation of the VLSI circuit, which, at that time, was unfamiliar technology. The regular and local connections reduce the input/output (I/O) demands, which reduce pin count of the circuit and minimize costly off chip communication. The processing elements are predetermined because they are to be implemented as a static, custom VLSI design. Pipelining is used to increase the throughput.

Many of the ideas behind systolic arrays are useful for reconfigurable devices. The main difference between systolic arrays and reconfigurable devices is that systolic arrays were aimed towards custom silicon, while reconfigurable devices are programmable.

1.3 Benefits of Reconfigurable Computing

Reconfigurable devices have used to implement a wide variety of applications. Early logic level reconfigurable systems were shown to be thousands of times faster than their current day microprocessor counterparts. Table 1 shows a number of different applications where reconfigurable devices is more efficient than a microprocessor.

Table 1: Comparison of the execution time of different applications on a reconfigurable devices and a microprocessor.

Algorithm	Reconfigurable System	Comparison CPU	Speedup
DNA Matching [31, 32]	SPLASH-2	SPARC 10, Cray2, 16K CM2	4300, 300, 200
RSA Crypto [5]	PAM	Alpha 150 MHz	17.8
SAT [33]	IKOS VirtualLogic SLI Emulator (16 Xilinx XC4013E)	Sun 5	100
Serpent Block Cipher [34]	Xilinx Virtex XCV1000	200 MHz Pentium Pro	18
Sieving for factoring large numbers [35]	Mojave	200 MHz UltaSparc	28
String Pattern Matching [36]	Xilinx XC6216	300 MHz Pentium II	29.9
LZ data compression [37]	Wildforce (4 Xilinx 4036XLA)	450 MHz Pentium II Xeon	30
Traveling Salesman [38]	SPLASH-2	125 MHz PA-RISC	4
Spec92 [26]	MIPS+RC	MIPS	1.12
Shape Adaptive Template Matching [39]	Virtex 1000E	1.4 GHz Pentium 4	7000
RC4 Key Search Engine [40]	Xilinx Virtex XCV1000-E	1.5 GHz Pentium 4	58

Reconfigurable devices have been used in a wide range of application in addition to these already shown in Table 1. Reconfigurable devices are used as emulator for neural networks [41-43] and continuous speech recognition systems [44]. They are used widely in image processing applications, including computer graphics radiosity [45] and medical image processing [46]. Various network routing tasks can be implemented efficiently on reconfigurable architectures [47-50]. Every year bring new applications of reconfigurable devices appear every year as witnessed by the large number of publications in conferences such as the Design Automation Conference (DAC), the International Conference on Computer Aided Design (ICCAD), the International Symposium on FPGAs (FPGA) and the International Symposium on Field-Programmable Custom Computing Machines. Browse through any of these proceedings and you will find many more applications being implemented on reconfigurable devices.

1.4   Summary

This section attempted to provide an ostensive definition for reconfigurable computing. We started by looking at some of the basic underlying concepts of reconfiguration. We talked about what meant to configure a device and looked at the relationship between programmability and reconfigurability. We provided a set of three characteristics that can be used to classify reconfigurable devices. Granularity is the abstraction level used to program a reconfigurable device. Coupling is the way the reconfigurable device is integrated into the overall system. Dimensionality is the communication scheme and physically arrangement for the processing elements of the reconfigurable architecture. Additionally, we discussed the benefits of reconfigurable computing across a wide variety of applications. The next section surveys a wide variety of reconfigurable architectures. We use the classification scheme to differentiate the different architectures. We start by describing devices with logic level granularity and continue until we have reached system level reconfigurable devices.

2.   RECONFIGURABLE ARCHITECTURES

You can trace the roots of reconfigurable systems to Gerald Estrin's work at UCLA in the 1960s. Estrin's "fixed plus variable structure computer" [51] consisted of a standard processor augmented by an array of reconfigurable hardware. His idea was well ahead of the technology at that time, so he was only able to make a crude approximation of his vision.

Interest in reconfigurable systems was renewed in the mid eighties with the emergence of programmable logic devices. These devices provided a large number of configurable logic units and programmable interconnect. Since that time, there has been much research and development in the field of reconfigurable systems. In this section, we highlight some of the important reconfigurable architectures that have emerged in the past 20 years. We start by looking at fine grain programmable logic devices that jumpstarted the field of reconfigurable computing. From there, we go on to profile some of the more coarse grain reconfigurable architectures that have emerged.

This section provides a glance at a number of different reconfigurable architectures. Research on reconfigurable architectures has become so popular, that it would be impossible to mention every reconfigurable system that has been proposed or developed. With that in mind, we attempted to find a spectrum of reconfigurable architectures that give good coverage of the research in this field.

2.1 Boolean Level Granularity

We begin the discussion of reconfigurable architectures by looking at programmable logic devices (PLDs). One can argue that these fine grain programmable devices started the revolution of reconfigurable computing. We start by looking at the basic components of PLD. Then, we continue on to the more complex PLDs and discuss some of the architectural features of these devices. Realizing that there is no way to cover this with much depth, we give only a limited description of FPGA and CPLD architectures. We refer the interested reader to Trimberger's book [52] for more specific details on these devices.

2.1.1 Programmable Logic Device (PLD)

A *programmable logic device* is a broad term that encompasses a range of integrated circuits that can be configured by the end user to implement complex logic functions. Much like the term "reconfigurable", the term "PLD" has been used (often in conflicting manners) by different people to mean a wide variety of things. We use a definition in a similar spirit to that of Brown and Rose [53], though we group programmable memory into PLDs for the convenience of presentation more than any other reason. Figure 5 shows our classification system for programmable logic devices. The remainder of this section will provide details on the different architectures of these logic devices.

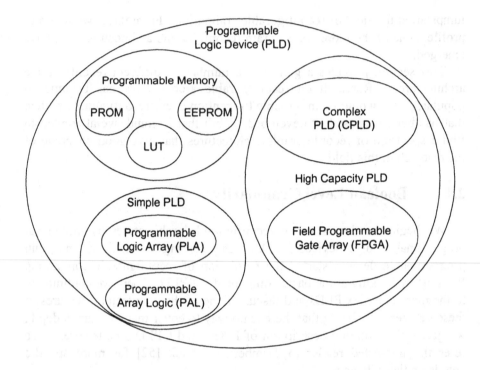

Figure 5 Venn diagram classification of programmable logic devices

Programmable memory is an array of memory elements that can be programmed in some manner after fabrication. The type of programming differs with the technology used to program the device. Some programmable memory works using fuses. Some use special floating gate structures to change the operation of transistors. Others have static RAM cells that change the bits that are stored in the programmable device. Each of these technologies has benefits and disadvantages. We start of discussion of programmable logic devices talking about different types of programmable memory.

Programmable read-only memory (PROM) is read-only memory (ROM) that can be modified once by a user. The PROM is constructed using a decoder and an array of AND or OR gates. The OR (AND) gates include a fusible link, which is used to program the PROM. A PROM is programmed using a machine that supplies an electrical current to specific cells in the ROM that effectively blows a fuse in them. The process is known as burning the PROM. This process leaves no margin for error; if you program the chip and find an error in your design, you must discard that particular PROM device and use another device.

Reprogrammable ROM chips emerged in the form of *erasable programmable read-only memory (EPROM)* or *electrically erasable programmable read-only memory (EEPROM)* technology. These are variants of PROM that can be programmed many times albeit through complex programming circuitry that often must reside off chip. The EPROM/EEPROM use floating gate technology on the AND or OR gate array. The gate can be set or reset to program the device. EPROM uses UV light to program the gate while EEPROM by applying a large bias on the gate in order to deposit/remove electrons on the floating gate.

Lookup tables (LUTs) employ static RAM cells as programming bits. Static RAM cells are built using two inverters and a couple of pass transistors (see Figure 6). Data is stored or read from the SRAM cell as long as the cell is powered. Without power, the SRAM cell loses its value, hence the "static" nature of SRAM. A LUT is an extremely generic computational component. It can compute "any" function; that is any n-input LUT can be used to compute any n-input function. A LUT requires 2^N bits to describe, but it can implement 2^{2^N} different functions. LUTs are limited to a small number of inputs due to the size of SRAM cells as a programming point. SRAM cells are larger than antifuses, EEPROM and EPROM.

SRAM cells provide numerous advantages. The configurations can be easily reprogrammed using on-chip circuitry. Therefore, it is easy to update LUTs for bug fixes or upgrades. SRAM devices are beneficial in the cases where many different configurations or modes of operations are used. For these reasons, we will see that most FPGAs consist of LUTs as their basic unit of computation.

Programmable memory can be used as a programmable computational unit by using the address lines as inputs and the data lines as outputs. Programmable memory with N address lines and M data lines is capable of computing any N input/M output function. However, logic functions are usually more efficiently realized in a network or multilevel fashion.

Programmable memory can be used as a microcontroller in the design of microprogrammable control units for data path circuits. Programmable memory uses a fully populated logic description of the function. This usually means that the programmable memory is over designed to implement the function at hand. Programmable memory is quite inefficient when implementing combinatorial circuits with lots of don't care terms.

Simple PLDs (SPLDs) are relatively small devices that focus on implementing logic functions in a two level, sum of product form. SPLDs are one of the earliest programmable logic devices and are still used as basic components in modern day CPLDs. SPLDs have a simple routing structure, which yield predictable delays.

a)

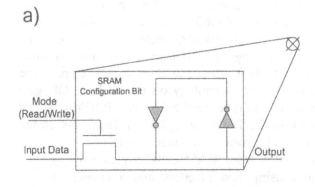

b)

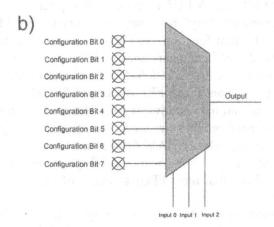

Figure 6 The basic structure of a LUT. Part a) shows the basic structure of an SRAM
configuration bit. Part b) is a 3 input LUT.

The *programmable logic array (PLA)* consists of a plane of AND gates
feed into a plane of OR gates. The functionality of the PLA is set when the
fuses between the inputs and the AND gates are blown in conjunction with
the fuses connecting the outputs of the AND gates and the inputs of the OR
gates. A PLA with N inputs and M outputs can implement any N input/M
output logic function. Figure 7 shows a 3 x 7 x 4 PLA. This PLA has 3
inputs (including their inverses), which feed into 7 AND gates. The outputs
of the AND gates are interconnected with the inputs of 4 OR gates in the
programmable OR array.

The PLA was introduced in the mid 1970s. When they were first
introduced, PLAs were expensive to manufacture and had poor performance
due to the two levels of programmability. The poor performance was mainly

due to the fuses, which have high capacitance and resistance. This made the interconnect delay extremely large.

The *programmable array logic (PAL)* is a special case of the PLA with the AND array being the only programmable array; the OR array of the PAL is fixed. The fixed OR array increases the performance compared to a PLA; however, a PAL is not capable of implementing any arbitrary N input/M output function.

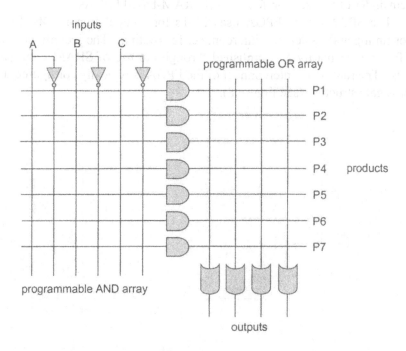

Figure 7 A 3 x 7 x 4 Programmable Logic Array (PLA)

The main problem with SPLDs is the large amount of overhead that the fuses add to the performance (delay, area, etc.) of the device. The PAL and PLA are one time programmable with fuses that are large and add much area and propagation delay to the circuit. Furthermore, while these devices are configurable, they are not reconfigurable since they can only be programmed once. Yet another problem with SPLDs is that they do not scale well. The architectures of these devices are restricted to two level implementations. Large input/output logic is often more efficiently implemented when using a multilevel logic implementation.

The *field programmable gate array (FPGA)* is potentially more efficient than programmable memory and SPLDs because FPGAs can implement multilevel logic functions. An FPGA is a collection of programmable gates embedded in a flexible interconnect network. They provide a scalable, user programmable alternative to SPLDs. FPGAs may use a variety of different types of programmable gates. They may use antifuses, EPROM or EEPROM as programming points. While many FPGA architectures use these technologies, most architectures use SRAM as a programming point. The SRAM makes the FPGA volatile, meaning it must be programmed every time that it is started up. Most importantly, SRAM allows for reconfiguration, which is the heart of reconfigurable computing. The remainder of this section focuses on SRAM-based FPGAs.

The SRAM-based FPGA uses LUTs for computational units, flip flops for timing and switchable interconnect for routing. The reconfigurability of FPGAs is primarily accomplished through the use of SRAM configuration bits. The routing or interconnect of the FPGA is typically configured using a pass gate structure (see Figure 8).

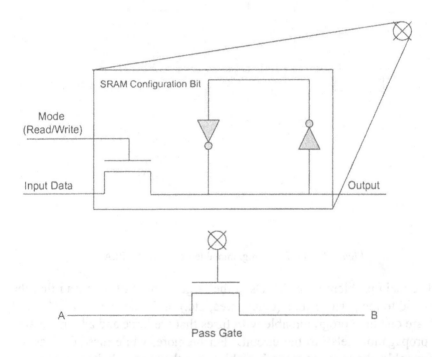

Figure 8 Configurable interconnect using a pass gate controlled by single SRAM configuration bit. If the SRAM bit is set, then A is electrically connected to B. Otherwise, A and B are not connected.

The LUTs typically have 4-5 inputs. Much empirical work on the best size LUTs along with other aspects of FPGA architecture has been done, largely by the researchers at the University of Toronto [54].

Modern FPGA architectures consist of arrays of *configurable logic blocks (CLBs)*. A CLB is a complex block consisting of LUTs, multiplexors and flip flops. The Xilinx 3000 series CLB [55] uses a relatively simple CLB; we will use this as a sample CLB architecture.

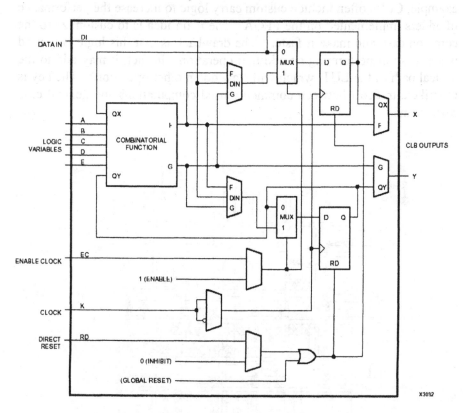

Figure 9 The Xilinx 3000 series configurable logic block (CLB) [55]

Referring to Figure 9, the CLB has five input logic variables (A, B, C, D, E) and two output variables (X, Y). It contains two D-type flip flops, which store the output variables X and Y. Combinatorial function generator has seven input variables – the CLB inputs (A, B, C, D, E) along with the CLB latch flip flop values (QX, QY). Internally, it is created using two 4-input LUTs. The combinatorial function generator has three modes of operation. It can implement any two independent 4-input functions, any one 5-input function or a limited number of 7-input functions. The different modes of operations have different requirements on the inputs, e.g. in the two independent 4-input functions, the variable A, B, C and D or E can be used as inputs, but both D and E can not appear in the same 4-input function.

More advanced FPGAs have more complex CLBs. For example, the CLBs in a Virtex II Pro consist of four separate slices, where each slice contains two CLBs [56]. Additionally, the CLBs contain custom logic for more efficient implementation of commonly used function units. For example, CLBs often include custom carry logic to increase the performance of adders implemented on the FPGA. The main idea is to customize to the common case and make it faster. The drawback is that this logic is unused when it is not performing an addition operation. In fact, it may add to the critical path of the LUT, which will slow down other operations. The key is to strike a tradeoff between commonly used computations and general case performance.

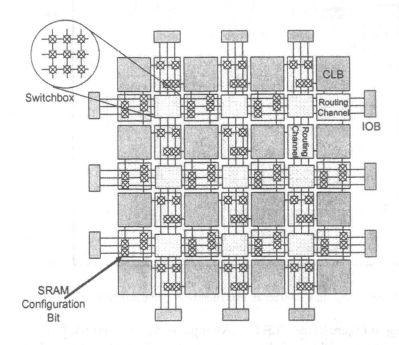

Figure 10 Generic island style (Toronto) FPGA model

Many FPGAs employs an island style routing architecture. The generic island style architecture (Figure 10) consists of two separate components for the routing architecture. The routing channel is a set of pass gates that allow provide programmable connections into and out of the CLB. The switchbox provides point to point connections between neighboring routing channels. Routing channels often contain long lines, which are used to span multiple CLBs in a row or column. The long lines create fast global connections between CLBs, which would otherwise have to pass through multiple, extremely slow switchboxes. The routing architecture is the major factor in both delay and area of the FPGA. Approximately 90% of the area of a typical FPGA is used for interconnect.

Take the Xilinx 4000 series as an example routing architecture. The 4000 series uses an island style architecture. Each CLB has twelve inputs that are connected to each of the routing channels adjacent to the CLB. The four CLB outputs are each connected to two of the four routing channels, one vertical channel and one horizontal channel. This increases the routability of the output signals. Not all of the inputs and outputs are normally used; in fact it is common that most will be unused.

The routing structure has three types of routing segments - single, double and long. Single length segment span only a signal routing channel before entering a switchbox. Double length segments allow for longer transmission with shorter propagation delays; they span two routing channels before connecting to a switchbox. Long lines cover half the length of the chip without entering a switchbox. They permit long distance communication to occur without excessive delays.

Island style architectures use relatively powerful functional blocks (CLBs) connected with versatile but expensive interconnect. Since the space taken by a CLB is a fraction of that used by the interconnect, it is worthwhile to create more powerful CLBs to reduce the amount of computational logic needed to program a circuit. An alternative approach is to make the interconnect far less versatile and far less expensive. This is the essence of the "cellular-style" [57] or mesh based architectures.

Triptych [58-60] uses a mesh based architecture. It involves a homogeneous array of very simple logic blocks with limited interconnect. Each individual cell is relatively simple and compact, which allows for high computational density. Logic blocks may act as interconnect to provide extra routing resources if more sophisticated routing is required. This approach simplifies the placement and routing process and offers density improvements in applications with regular structure. Montage [61] is an asynchronous mesh based architecture. The Montage architecture consists entirely of asynchronous logic while maintaining the same general architecture of Triptych.

One of the main features of an FPGA is its complete reconfigurability. Both the interconnections (routing) and the computation (logic functions) can be reprogrammed. Thus, the same device can be reused to realize completely different logic functions. The reconfigurability in FPGAs is primarily accomplished through the use of the SRAM configuration bits. The routing or interconnect of the FPGA is typically configured using a pass gate structure. The computation is configured through SRAM configuration bits on multiplexors or LUTs.

The SRAM programming bits are distributed across the entire FPGA. The large internal configuration memory makes it hard to get the configuration data into the device. Essentially, it is feed in serially, which causes a huge overhead for reconfiguration. This has been partially ameliorated through configuration caching and compression [62-66].

The FPGA has two modes of operation. The download mode brings configuration into the device memory and the configured mode runs the device as specified by the configuration. The synthesis flow of the FPGA largely mimics that of the hardware synthesis design flow. Therefore, compiling a program on the device is very slow. In addition, configuration time (download mode) is lengthy due to the low level of configuration granularity.

Some researchers have looked at architectural techniques to reduce the configuration time. Multiple contexts and partial reconfiguration are two particularly effective architectural features for reducing configuration time in an FPGA.

Multiple context FPGAs provide a small number of programming bits at each programming point. Each of the different bits corresponds to a different context. Because the context bits are already on chip, the device can quickly switch between these different contexts. However, the download time of the FPGA is increased because we effectively multiplying the number of programming bits of the FPGA by the number of contexts.

The DPGA [67-69] was one of the first attempts at a multi-context FPGA. The DPGA uses a bit level (logic level) architecture with multiple on-chip instructions per compute element. The DPGA has space allocated on the chip to hold the configurations for each gate and switch. Each LUT and multiplexor is associated with a small local memory (see Figure 11). A broadcast identifier decides between the particular contexts, i.e. the particular configuration bit, at any given point of time. Of course, these multiple contexts are beneficial only to the extent that even with the additional space on chip, there is a net increase in device utilization.

In applications where the throughput is not as important a consideration as the distinct functionality provided, the different stages in the pipeline are treated as different contexts. Levelized logic [68] is a technique for automatic temporal pipelining of existing circuit netlists. The basic idea is to use slack in the netlist to equalize the context sizes and hence reduce capacity usage.

Typically FPGAs are single context devices; the resources on the chip allow it to perform a single function throughout the application. At many points during the execution of the program, there will be certain portions on the chip that will not be used at all or will be used very infrequently. If we could reuse these portions for other tasks that are more in demand, we could improve device utilization and also in some cases the performance. Multiple context devices are programmable arrays that allow the strategic reuse of these limited resources. In other words, multiple context devices exploit the temporal aspect of resource capacity (gates and interconnect) to improve the efficiency.

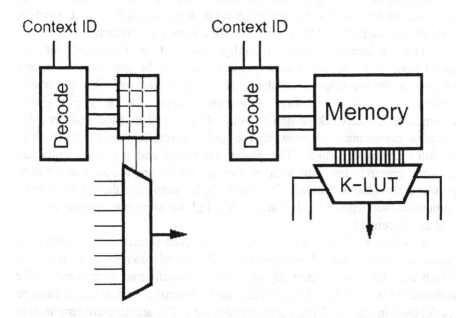

Figure 11 Multiple context computing elements. Each programming bit is replaced with a small memory of programming bits, which correspond to the different contexts. [68]

Partial reconfiguration provides another method to ameliorate large configuration times for FPGAs. The DISC – dynamic instruction set compiler - project [70] was one of the first academic attempts at partial reconfiguration.

The overriding goal of the DISC project was to compile C code fragments to assembly language for a custom FPGA-based processor. DISC works on the idea of the partial reconfiguration and run time reconfigurability. DISC treats instructions as removable modules paged in and out through partial reconfiguration as demanded by the application. The instructions occupy the FPGA space only when required, and the FPGA resources can be reused to implement an arbitrary number of instructions.

The DISC architecture consists of a single National Semiconductor CLAy31 FPGA coupled to an external RAM. The processor consists of a global controller and custom instruction modules. The global controller provides the circuitry for operating and monitoring the global resources such as the external RAM, I/O, the internal communication network and the global state. It consumes about 1/6th the area of the chip.

The major idea behind partial reconfiguration is the fact that reducing the size of the hardware to configure significantly reduces the configuration bit stream. The project claims the potential of 3 to 60 times reduction in configuration times compared to that of configuring an entire FPGA. In a sense, DISC acts like an instruction cache. When a new instruction arrives and there is no space in the reconfigurable array to implement it, existing instructions are removed to make space for this new instruction.

When performing partial reconfiguration, it is important for the instructions to be designed in a way that they can be implemented at any location on the reconfigurable fabric. The DISC project calls this relocatable hardware. This flexibility requirement calls for a good global communication network, which is available at every location and provides adequate communication between the global controller and the instruction modules at any location. The global controller and the communication network remain in the same location throughout the application execution to preserve the global context. The modules lie perpendicular to the global communication signals for full access of all global signals regardless of their vertical placement.

The Xilinx 6200 series [71] was a short lived commercial attempted at partial reconfiguration. It incorporates a RAM-style configuration interface which permits it to be partially or wholly reconfigured at runtime. The architecture is mesh based rather than the traditional island style architecture model used in most of Xilinx previous devices. The architecture emphasizes locality with 4 by 4 grouping of CLBs coupled with fast local interconnect. The 6200 series was perhaps a little bit ahead of its time; it had little commercial success, but inspired much interesting research [62, 64, 72-85]. Most modern high performance FPGAs use some sort of partial reconfiguration.

Complex programmable logic devices (CPLDs) are arrays of SPLDs with programmable interconnect architecture. CPLDs were pioneered by Altera, which were originally called classic EPLDs. Initially, CPLDs were viewed as an extension of SPLDs. The idea was to increase the size of SPLDs in order to handle larger designs. However, just increasing the AND/OR planes of SPLDs creates a structure that grows too quickly with the number of inputs. The solution was to create arrays limited input/output SPLDs on a single chip.

We illustrate some of the features of CPLD architectures using the Altera 7000 series architecture. The 7000 series is an array of SPLDs. It uses a two level hierarchical architecture. The architecture consists of two components the *logic array block (LAB)* and a *programmable interconnect array (PIA)* at the highest level of the hierarchy.

The LABs have a complex SPLD-like structure. Each LAB is composed of 16 macrocells. A macrocell is essentially a five product term PAL.

Macrocells can be configured as sequential or combinatorial logic. There are 16 macrocells per LAB. The LABs are interconnected through the PIA. The PIA is a fully connected switchbox. It can connect any LAB input/output to any other LAB.

2.2 Instruction Level Granularity

In this section, we look at reconfigurable architectures that have instruction level granularity. These architectures often consist of some sort of reconfigurable functional unit (RFU) coupled in some manner with a controller. Most often the controller is a general purpose microarchitecture, e.g. RISC processor. When coupling with general purpose processor (GPP), the GPP acts as a master while RFU acts as a coprocessor. The GPP is in charge of controlling the RFU and executing program code that can not be effectively speed up using the RFU. Most often, this corresponds to control intensive code or code that exhibits a limited amount of parallelism.

The coupling of these systems is a main point of differentiation between different architectures. In a closely coupled system, the RFU plays the role of a customizable functional unit. A loosely coupled system might use a PCI bus or even a network interface. The coupling plays a large role in determining the amount of code that can be executed by the RFU. If the system is tightly coupled, much more code can be implemented on the RFU since the time to transfer the configuration bits and data to the RFU is small (on the order of 10s of cycles). However, a tightly coupled system is limited in size (and therefore the amount of parallelism) because it occupies expensive on chip resources. A loosely coupled system can afford to have much larger computational area (often arrays of FPGAs), making it better for applications with large amounts of parallelism. However, there is a large latency between the CPU and RFU making that the dominate factor in choosing what code can and can not be executed on the RFU.

We continue our survey of reconfigurable architectures by looking at a series of reconfigurable devices that couple with a GPP. We start by looking at tightly coupled architectures and continue to look at different reconfigurable microarchitectures that are more loosely coupled.

We start our discussion with the PRISC project [27, 86]. PRISC incorporates hardware programmable resources into GPP microarchitecture to improve the performance of general purpose applications. The hardware programmable unit is called a *programmable functional unit (PFU)*, which corresponds to a RFU in our lingo. The PFU is automatically configured each time a new functionality has to be implemented by it.

The architecture consists of a RISC microprocessor with a single PFU. The conventional set of RISC instructions are augmented with application specific instructions that are to be implemented in the PFU. The PRISC datapath is shown in Figure 12. The PFU is tightly integrated into the datapath of the GPP. The PFU acts much like a functional unit; it takes two input operands from the register file and can write one operand back into the register file. The tight coupling restricts the amount of processing that can be done using the PFU and the area of the PFU. It should be on the same scale as the other datapath functional units. The PFU is essentially a two dimensional array of LUTs where the data flows through three alternating layers of interconnect and rows of LUTs.

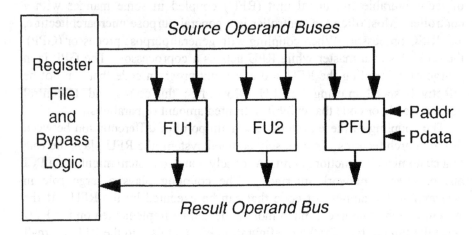

Figure 12 The datapath of the PRISC architecture [27].

Compilation to the PRISC architecture is done in the same manner as compiling to any other GPP microarchitecture, except for \hardware extraction. This step identifies sets of sequential instructions (these are a list of candidate PFU instructions) that can be implemented on the PFU. These instructions are given to hardware synthesis tools to convert them into a netlist of LUTs.

Most of the instructions implemented on the PFU are combinational operations and byte-wise add/subtract type of operations. The reason why other operations are not considered for implementation on PFU is because they will be executed much faster on the RISC processor, which has dedicated functional units to execute them. The algorithm for hardware extraction does a backward traversal on the graph until it finds a *non-PFU* instruction. It stops here, and checks if the PFU has enough resources to implement the function. If not, it gives feedback to the extraction routine to reduce the size of the function.

The Chimaera project [28] extends the PRISC project by enhancing the architecture of the RFU. The Chimaera RFU is reconfigurable array (RA) consisting of logic blocks with two 2-input LUTs and one 3-input LUT. The logic blocks are purely combinatorial, meaning that the RFU cannot hold state. The array is arranged into a number of rows, where each row has 32 logic blocks. The each row can draw from a subset of the registers from the register bank. The prototype allowed access to registers R0-R8 through a shadow register file [87]. The logic blocks and routing are designed to facilitate fast addition, subtraction and comparison.

Operation performed on the RFU look like operations between registers in the register file. The RA acts as a cache for RFU operations. The RA can hold many different RFU operations. The operations span must span multiples of rows; a row of logic blocks is the atomic unit of reconfiguration. The configuration control and caching unit (CCCU) is responsible for loading and caching the configuration data corresponding to the RFU operations that reside in the RA. The CCCU supports partial reconfiguration of the RA. The execution control unit (ECU) decodes the instruction stream and directs the execution of RFU.

The RA can reduce the execution time of the program by implementing dependent instructions as an RFU operation. Also, it can speculatively execute branches to reduce dynamic branch count and exploit sub-word parallelism. Furthermore, using the RFU can reduce resource contention (instruction issue bandwidth, write back bandwidth or functional units). The Chimaera project includes a C compiler built on top of GCC 2.6.3. The primary additional optimizations are instruction combination, control localization and SIMD within a register. Instruction combination finds sequences of instructions in single control flow. Control localization performs optimizing/reduction techniques on branches to easily create an RFU operation. Finally, SIMD Within a register (SWAR) exploits sub-word parallelism where ever possible.

The OneChip project [88] uses RFUs with sequential components i.e. the RFU holds state. This allows the RFU to implement any combinatorial or sequential circuit, subject to size and speed. OneChip still uses a tightly coupled approach of integrating reconfigurable logic into a processor data path. The OneChip RFU is implemented in parallel with the other basic elements of the data path. The RFU is controlled through instruction stream of the microprocessor. The RFU works on register to register operations. Therefore, it works best on operations that have limited input but do a lot of computation.

The Garp project [26, 29] couples a MIPS-II clone and 32 by 24 reconfigurable array of LUT-based 2 bit processing elements. The reconfigurable array is primarily a mesh architecture. The basic unit of configurations is a row (32 processing elements). The host processor has additional configuration instructions used to program and control the reconfigurable array. The host processor and RA use the same memory and the RA can initiate memory access, however the access are restricted to flow through the central 16 columns.

The Garp RFU is more loosely coupled than the previous discussed architectures. The RFU acts as a slave to the GPP microprocessor. It is designed to fit into a typical computing environment supporting structured code threads, libraries, context switching and virtual memory. This looser coupling means that the RFU can be large; hence a larger amount of code can be implemented on it. Assembly instructions are used to control the RFU (e.g. mtga, mfga). The RFU can access memory through the cache. Additionally, the reconfigurable array has an internal configuration cache for context switching. The RFU incurs no overhead during a context switch.

The Garp project looks for hyperblocks [89] to implement on the RFU. Hyperblocks are intra-function control flow sequences with a single entry and one or more exits. Hyperblocks consist of many instructions (on the order of 10 – 100 instructions). The GPP stalls and transfers control of the program to the RFU when the RFU is executing.

The Garp processor borders on an architectural level granularity. The RFU is large enough that it can begin to handle function level operations.

The PRISM project [25, 90] looked at a loosely coupled system consisting of a general purpose processor (Motorola 68010 processor) with reconfigurable component consisting of several FPGAs (Xilinx 3090 FPGA). The GPP controls the execution and uses the RFU as a coprocessor. When the GPP decides to use the RFU, it will load the data into the RFU, stall until the RFU has completed computation, and store the results of the RFU operation back into the GPP and continue execution. The GPP and RFU are connected through a 64 bit bus. The configurations for the RFU are stored in memory. The time to load an adaptive element was around 30 ns for PRISM-II.

The PRISM project looked at automating the selection of the RFU operations. It used C syntax to describe functions as a control data flow graph (CDFG). A function could either be mapped onto the GPP or the RFU. The PRISM compiler automatically generates the configuration bitstream to program the RFU. However, the user is responsible for deciding the mapping of the function.

The next section will continue our survey of reconfigurable devices. It will describe reconfigurable architectures that perform increasingly complex operations and more coarse grain coupling of a variety of programmable components.

2.3   Function Level Granularity

The reconfigurable devices in the previous section all coupled a general purpose processor with a reconfigurable functional unit (RFU). In each case, the RFU was implemented using a logic level programmable architecture. The logic level granularity of the RFU resulted in a large number of programming bits for the RFU, meaning that it was slow to program. Additionally, the tight coupling required that the RFU be small. Therefore, number of operations that can be scheduled on the RFU are limited.

Function level granularity is a coarser grain reconfigurable device, where the reconfigurable architecture consists of programmable components that can implement complex multicycle operations. The components of these architectures are microarchitectural and architectural level elements like ALUs, registers and other data path elements. We start our survey of function level granularity with reconfigurable architectures that use fine grain function level granularity and continue on to more architectures with increasingly course grain function level granularity.

The PADDI (Programmable Arithmetic Devices for high speed Digital signal processing) project [91-93] is a reconfigurable system aimed at high speed computation of DSP applications. PADDI is an interconnection of numerous simple processing elements in a MIMD fashion. The PADDI architecture consists of 48 simple processing elements called "nanoprocessors". Each nanoprocessor consists of six general purpose registers, three scratch pad registers, an eight element instruction memory and a simple control unit. PADDI has 16-bit functional units with an 8-word deep instruction memory per processing element. A chip-wide instruction pointer is broadcast on a cycle-by-cycle basis giving PADDI a distinctly VLIW control structure. PADDI-2 also supports 8 distinct instructions per processing element but dispenses with the global instruction pointer, implementing a dataflow-MIMD control structure instead.

The chip is organized into 12 clusters of 4 nanoprocessors. A two-level interconnect scheme is used, which allows one cycle communication between any two nanoprocessor. PADDI uses the Silage language [94, 95] to specify the DSP application. Silage is a good language for specifying data flow graphs. The PADDI architecture has shown impress performance on a variety of DSP applications including filtering, discrete cosine transform and Viterbi decoding.

PipeRench [96-101] is a reconfigurable fabric – an interconnected network of configurable logic and storage elements. PipeRench is designed to efficiently handle computations. The PipeRench RFU is partitioned into highly parameterized "stripes" that they can be easily and quickly reconfigured. Each "stripe" can produce a datapath and can be likened to a pipeline. The stripe performs computationally complex arithmetic operations. Each stripe is composed of a number of processing elements, which contain an ALU, control and carry bits and a barrel shifter. Registers separate the stripes creating a pipeline.

PipeRench is programmed using the Cached Virtual Hardware Assembly (CHASM) language. It provides straight-forward programming and compilation into PipeRench. CHASM is a dataflow intermediate language – a single-assignment language with C operators where programmers to specify the bit width of variables.

RaPiD [102-106] is a general coarse-grained reconfigurable architecture that allows the user to construct custom application-specific architectures in a run-time configurable way. RaPiD uses coarse grained computational units and a one dimensional interconnect structure.

The RaPiD architecture consists of a set of application-specific function units, such as ALUs, multipliers, shifters and bit-configurable operators. The architecture includes a set of application-specific memory elements, including registers and small data memories. A set of input and output data ports interface the datapath to external memory and streaming data interfaces. An interconnection data network connects the function units, memory elements and data ports together using a combination of configurable and dynamically controlled multiplexers. A sequencer generates "instructions" that control the operation of the RaPiD datapath. An interconnection control network generates the individual control signals based on the instructions and status signals generated by the function units.

MATRIX (Multiple Alu archiTecture with Reconfigurable Interconnect eXperiment) [67, 107] is composed of an array of identical, 8-bit functional units overlayed with a configurable network. Each functional unit contains a 256 x 8-bit memory, an 8-bit ALU and multiply unit, and reduction control logic including a 20 x 8 NOR plane. The network is hierarchical supporting three levels of interconnect. Functional unit port inputs and non-local network lines can be statically configured or dynamically switched [107].

RAW [108, 109] is a scalable, course grained array of processors that efficiently use space, utilize current hardware design technology and make more and better compile time decisions. The RAW processor was designed as a scalable ISA that provides a parallel software interface to the gate, wire and pin resources of the chip. It allows the programmer to specify communication primitives to exacerbate the increasing wire delay problem.

The RAW architecture is a mesh of tiles; each tile is a stripped down MIPS instruction set processor with 32 KB instruction memory and 32 KB data memory. The tiles communicate via registers and buffered in FIFOs. The communication may occur statically through a static switch processor that implements simple instruction set for moves, branches, jumps, and nops.

Also, a dynamic router supports communication which cannot be statically determined by compiler

Programs are mapped onto tiles using compiler and runtime support. Programs execute both in time and space; the structure of tiles for a given program matters. The compilation process starts by dividing a program into basic units. Then, the compiler determines inherent parallelism and generates multiple instruction streams. Finally, communication commands are inserted in the program and each portion of the program is mapped to a physical tile. Runtime software is used to ensure proper configuration and communication.

NAPA (National Adaptive Processing Architecture) [110] uses a number of processing units tied together using a wiring network. Each processing units consists of a fixed instruction processor (FIP) and an adaptive logic processor. The FIP is general purpose scalar processor (32 bit RISC microprocessor) used to execute the software-like portion of the application. The ALP is composed of core cells, which contain a D flip flop, 16 bit configuration register and a three input, two output logic cell, essentially a logic level reconfigurable device. The ALP is used to implement the hardware-like portion of the application.

NAPA is programmed using a special language called Napa-C [111, 112]. It uses pragma statements to specify the hardware/software partitioning. Based on the user partitioning, the NAPA compiler generates the C program for the FIP and the configuration to control the ALP. Additionally, the compiler inserts code into the FIP to control the execution and configuration of the ALP.

2.4 System Level Granularity

Reconfigurable devices with system level granularity consist of large computational units. Often, the computational units are general purpose processors, RAM blocks and FPGAs. The computational units are physically arranged in some manner, often in an array with different types of communication schemes. These systems have a large amount of processing power. They are often integrated into high performance computing systems as boards connected to the PCI bus. These boards act as processing cards in a similar manner to graphics and sound cards in current personal computers. Most often, these devices are used for emulation of VLSI circuits.

Programmable active memory (PAM) [5, 113-116] devices were designed to act like an customizable ASIC. They were envisioned to sit on some high speed bus of a host processor, like a RAM memory module. The processor can write into and read from PAM, just like it would communicate with RAM. PAM possesses data between write and read instructions, which make it an active memory. A memory write is sending configuration information and input data to the PAM component and a memory read is bringing the results of the computation back into the host processor. The specific processing is determined by the contents of the bitstream device, which can be changed by the host processor in a matter of milliseconds. Hence the device is programmable.

The speed critical parts of an algorithm are loaded onto PAM. The coding process was quite laborious and slow, but still much quicker than hardware design iterations. The XACT Design Editor is used to create data in the PAM model.

Though PAM was a virtual device, several physical realizations of PAMs e.g. DECPeRLe-1 and SPLAM, were created. These devices gave impressive performance in a wide variety of applications including crptyography, molecular biology, neural networks and high energy physics.

The DECPeRLc-1 has a two dimensional layout of Xilinx XC3090 FPGAs. Each FPGA is connected local to its four neighbors. The system also contains 4 megabytes of cache. The cache consists of four independent 32 bit wide SRAM banks. The device has four 32 bit wide external connectors, which are used to communicate to the outside processors.

A similar project with a slightly different focus was SPLASH [24]. Instead of the two-dimensional array architecture used by PeRLe, the SPLASH developers opted for a linear arrangement designed specifically for one-dimensional systolic problems. The communication architecture stresses neighbor to neighbor communications. While this architecture is less general than Perle, it performed extremely well in its application of DNA sequence matching; SPLASH was two orders of magnitude faster than Cray supercomputers.

2.5   Hybrid Reconfigurable Systems

Hybrid reconfigurable systems contain some kind of computational unit, e.g., ALUs, Intellectual Property units (IPs) or even traditional general-purpose processors, embedded into a reconfigurable fabric.

One type of hybrid reconfigurable architecture embeds reconfigurable cores as a coprocessor to a general-purpose microprocessor e.g. Garp [26] and Chimaera [28]. Another direction of new architectures considers integration of highly optimized hard cores and hardwired blocks with reconfigurable fabric. The main goal here is to utilize the optimized blocks to improve the system performance. Such programmable devices are targeted for a specific *context* – a class of similar applications, such as DSP, data communications (Xilinx Platform Series) or networking (Lucent's ORCA). The embedded fixed blocks are tailored for the critical operations common to the application class. In essence, the programmable logic is supported with the high-density high-performance cores. The cores can be applied at various levels, such as the functional block level, e.g., Fast Fourier Transform (FFT) units, or at the level of basic arithmetic operations (multipliers).

Presently, a context-specific architecture is painstakingly developed by hand. The *Strategically Programmable System (SPS)* [117, 118] explores an automated framework, where a systematic method generates context-specific programmable architectures.

The basic building blocks of the SPS architecture are parameterized functional blocks called *versatile parameterizable blocks (VPBs)*. They are pre-placed within a fully reconfigurable fabric. When implementing an application, operations can be performed on the VPBs or mapped onto the fully reconfigurable portion of the chip. An instance of our SPS architecture is generated for a given set of applications (specified by C or Fortran code). The functionality of the VPBs is tailored towards implementing those applications efficiently. The VPBs are customized and fixed on the chip; they do not require configuration, hence there are considerably less configuration bits to program as compared to the implementation of the same design on a traditional FPGA.

The motivation is to automate the process of developing hybrid reconfigurable architectures that target a set of applications. These architectures would contain VPBs that specially suit the needs of the particular family of applications. Yet, the adaptable nature of our architecture should not be severely restricted. The SPS remains flexible enough to implement a very broad range of applications due to the reconfigurable resources. These powerful features help the architecture maintain its tight relation to its predecessors, traditional FPGAs. At the same time the SPS is forming one of the first efforts in the direction of context-specific programmable devices.

In general, there are two aspects of the SPS system. The first area involves generating a context-specific architecture given a set of target applications. Once there is a context-specific architecture, one must also be able to map any application to the architecture.

The main components of SPS are the Versatile Parameterizable Blocks (VPBs). The VPBs are embedded in a sea of fine-grain programmable logic blocks. We consider a lookup table (LUT) based logic blocks commonly referred to as combinatorial logic blocks (CLBs), though it is possible to envision other types of fine-grain logic blocks, e.g. PLA-based blocks.

Essentially, VPBs are hard-wired ASIC blocks that perform a complex function. Since the VPB is fixed resource, it requires little reconfiguration time when switching the functionality of the chip. By functionality, we mean the application of the chip can change entirely, e.g. from image detection to image restoration, or part of the application can change, e.g. a different image detection algorithm. Therefore, SPS is not limited by large reconfiguration times like current FPGAs. But, the system must strike a balance between flexibility and reconfiguration time. The system should not consist mainly of VPBs, as it will not be able to handle a wide range of functionality.

There is a considerable range of functionality for the VPBs. It ranges from high-level, intensive tasks like FFT to a "simple" arithmetic task like addition or multiplication. Obviously, there is a large range of complexity between these two extremes. Since we are automating the architecture generation process, we wish to extract common functionality for the given context (set of applications). The functionality should be as complex as possible while still serving a purpose to the applications in the context. An extremely complex function that is used frequently in only one application of the context would be wasted space when another application is performed on that architecture.

The past decade has brought about extensive research as to the architecture of FPGAs. As we previously discussed, many researchers have spent copious amounts of time analyzing the size and components of the LUT and the routing architecture. Instead of developing a new FPGA architecture, the SPS leverages the abundant body of FPGA architecture knowledge for our system. Embedding the VPBs into the programmable logic is the most important task for the SPS architecture generation.

The architecture formation phase and the architecture configuration phase are the two major parts of the SPS framework. The SPS framework is summarized in Figure 13.

Architecture formation is described as making the decision on the versatile programmable blocks to place on the SPS chip along with the placement of fine-grain reconfigurable portion and memory elements, given an application or class of applications. In this phase, SPS architecture is customized from scratch given certain directives. This process requires a detailed description of the target application as an input to the formation process. A tool will analyze these directives and generate the resulting architecture. Again the relative placement of these blocks on the SPS chip along with the memory blocks and the fully reconfigurable portions need to be done by the architecture formation tool.

Unlike a conventional fine-grain reconfigurable architecture, a uniform distribution of configurable logic blocks does not exist on the SPS. Hence for an efficient use of the chip area as well as high performance, the placement of VPBs on the chip and the distribution of configurable logic block arrays and memory arrays among those are critical. The routing architecture supports such hybrid architecture is equally important and requires special consideration. If the routing architecture cannot provide sufficient routing resources between the VPBs and the configurable blocks, the hardware resources will be wasted. The type and number of routing tracks and switches need to be decided such that the resulting routing architecture can support this novel architecture most efficiently. The most important issues here are the routability of the architecture and the delay in the connections.

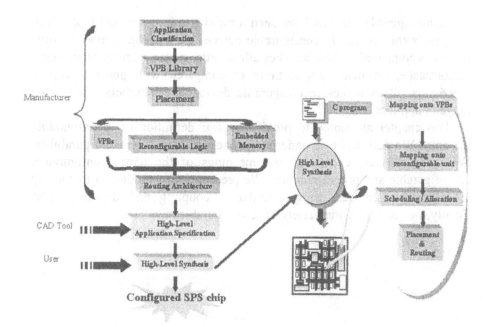

Figure 13 A detailed flow for configuring a Strategically Programmable System.

Another case that we are considering in our work is mapping an application onto a given architecture. At this point we need a compiler tailored for our SPS architecture. The SPS compiler is responsible for three major tasks:

The compiler has to identify the operations, groups of operations or even functions in the given description of the input algorithm that are going to be performed by the fixed blocks. These portions will be mapped onto the VPBs and the information regarding the setting of the parameters of the VPBs will be sent to the SPS chip. Secondly, the SPS compiler has to decide how to use the fully reconfigurable portion of the SPS. Based on the information on the available resources on the chip and the architecture, mapping of suitable functions on the fine-grain reconfigurable logic will be performed. Combining these two tasks the compiler will generate the scheduling of selected operations on either type of logic. Finally, memory and register allocation need to be done. An efficient utilization of the available RAM blocks on the SPS has to be realized.

3.   SUMMARY

Reconfigurable systems have seen a rapid rise in usage and research in the past twenty years. Reconfigurable devices are a cheaper, more flexible device as compared to ASICs. They allow better customization, hence better performance, on many applications as compared with general purpose processors. This makes reconfigurable devices are beneficial for a wide range of applications.

This chapter attempted to provide a clear definition of reconfigurable systems. We defined basic underlying concepts related to reconfigurability. We compared and contrasted the meanings of the terms configurable, reconfigurable and programmable. We presented a methodology to classify reconfigurable devices using granularity, coupling and dimensionality. Finally, we survey a wide variety of reconfigurable devices.

Chapter 4

MODELS OF COMPUTATION

1. INTRODUCTION

A *model of computation (MOC)* is a conceptual framework that allows a design to be specified, reasoned about, synthesized and tested. It is an unambiguous formalism that allows design specification, verification, optimization and synthesis [119]. More specifically, a MOC defines a set of computational components as well as the "laws of physics" that govern the interaction between the components [120].

Designers use different models of computation to abstract away unimportant aspects of the design. These system properties are considered to be either irrelevant at the current time or just not important in the design of the system. The specific model of computation has mathematical properties that allow the designer to formally prove properties about certain aspects of the design without going through the expensive task of formal verification or detailed simulation. For example, certain MOCs can be proven to be deadlock free.

We focus on models of computation that are used in digital systems – in particular embedded and reconfigurable systems. In this respect there are four general goals for a model of computation. 1) The MOC should allow a designer to capture the functionality of the system i.e. allow for *system specification*. 2) It must allow verification and/or simulation in order to verify the correctness of the specification. 3) The MOC should be synthesizable. 4) The MOC should enable efficient and effective estimation of various properties of the system. We will now explain each of these goals in more detail.

System specification is perhaps the most important part of system design. For without the ability to specify the system one wishes to design, the other tasks are frivolous. The language used to specify the system must be expressive enough to allow the designer to encode the system constraints and functionality. This is especially important for embedded systems. Embedded systems are often used in critical real-time applications (e.g. braking system for a car). They rely on hard reaction times for different components on the system. If we could not specify these hard constraints using a particular language, then that language is useless for our task. At the same time, a language with too much expressive power may allow the designer too much freedom, effectively crippling the amount synthesis, estimation and verification that a system can achieve. There is a delicate tradeoff between the expressiveness of a language and the three remaining goals of a model of computation.

The second goal for a MOC deals with the ability to reason about the correctness of the design as stated in the specification. There are two methods commonly used to validate the specification of the system with the actual functionality of the system – formal verification and simulation (testing). *Formal verification* establishes the properties of the system based on logical arguments or transformations. This involves formal specification of the requirement, formal modeling of the implementation, and precise rules of inference to prove that the implementation satisfies the specification. *Simulation* involves producing a set of inputs to the system (test pattern generation), applying them to the system and observing the output behavior. The resulting behavior should match the behavior as stated in the initial specification.

In general, simulation is easier than formal verification, but simulation cannot exactly verify the correctness of the system – something that formal verification can achieve. The model of computation can enhance or detract from the ease of verification. A MOC that allows transformations that can provably preserve the behavior of the system is extremely useful for formal verification. Yet, often these transformations limit the ability to produce an optimized final circuit. Once again, there is a tradeoff; this time it is between ease of verification and the ability to perform synthesis.

As defined in a previous section, synthesis is the transformation from a behavioral representation (at any level of abstraction) to a structural representation of the same level of abstraction. The model of computation is behavioral and is defined at a particular level of abstraction. In addition, it defines the types of optimizations that can be done in order to create a better system. The inherent properties of the MOC will determine the difficulty of the behavioral to structural transformation.

Estimation is the ability to accurately model different properties of a final implementation of the system before the system is fully realized. Estimation is used to guide different optimizations and transformations in order to create a system that is "better" with regard to some set of properties. For example, we may wish to make a system that has low power dissipation. We could perform different transformations and estimate the amount of power that the system will dissipate before and after each transformation. If the estimation is inaccurate, the transformations may have the opposite effect; they may create a system that dissipates excessive power. Estimation is extremely important to guide the synthesis process. Unfortunately, there is an inverse relationship between the level of abstraction and the ability to provide accurate estimation. The more abstract the model, the less accurate the estimation. Yet, an abstract model allows for a larger design space exploration allowing the possibility for a better final design.

As an example of the various tradeoffs between different MOCs, let us consider the finite state machine (FSM) and the Turing machine as two different ways to model a problem. There are many well-know methods of synthesis from the behavioral description given by the finite state machine (FSM) to the structural implementation. One could implement a FSM using a PLA or a ROM. Both of these structural representations are well defined and can be fully automated given any FSM description. In addition, we can easily verify that the structural representation is accurate according to the initial specification. Also, it will be fairly easy to estimate the various properties of the FSM implementation.

On the other hand, the synthesis of a specification using a Turing Machine as an MOC is not so deterministic and far from automatic. Due to the expressive power of the Turing machine, the structural realization of the specification is hard to optimize. We could use some generic processing unit (e.g. a microprocessor) that realizes the functions of a Turing machine, but this will often be overkill; it will often be too general for the specific task we wish to implement. It is hard to verify that this processing unit will yield a correct realization of the specification - this is akin to verifying the correct functionality of the microprocessor. In addition, the estimation using the Turing machine is arguably harder as opposed to the FSM. We have to account for and model the various components of the Turing machine. Assuming that we use a microprocessor to implement our Turing machine, we must understand the functionality of the different parts of the microprocessor, e.g. caching policy, instruction dispatching, branch prediction mechanism, etc.

Imagine that we modeled the same problem using a Turing Machine and a FSM. The synthesis of the problem using the FSM as a model would be efficient. Whereas it would be difficult to understand the simple nature of the synthesis problem when it is modeled using a Turing machine. It may be possible to surmise that the Turing machine model can be simplified to a FSM, but this is a difficult problem in itself. So it seems that we should always model a problem with an FSM. But then, we have limited ability of expressiveness. Imagine modeling a microprocessor with a FSM. It may be theoretically possible, but the sheer number of states would make the FSM specification impossible to manage.

In the next section, we will discuss the various properties that characterize a model of computation. We will describe the properties of a model of computation and attempt to present the terminology to allow a comparison between different models of computation. The following section will describe some of the classic models of computation and show how they differ from one another through their handling of different properties. Then, we will give an overview of many of the popular MOCs used to design embedded and reconfigurable systems.

2.   PROPERTIES OF MODELS OF COMPUTATION

It is hard to provide an all-encompassing method to compare models of computation. Most often, the application that is being modeled will dictate the MOC that is used. Different models have different properties that allow them to describe the some aspects of a system/application better than others. Nevertheless, it is useful to provide a high-level comparison of the different models so that a designer can understand the benefits and drawbacks for using a specific MOC.

There have been several attempts to compare different models of computation [121-123]. We will mainly draw from the terminology as described by Lee and Sangiovanni-Vincentelli as they focus on MOCs for embedded systems. Before we start comparing the different models, we wish to formalize some terminology to aid the process of comparison.

A *language* is set of symbols, rules for combining the symbols (*syntax*) and rules for interpreting a combination of symbols (*semantics*). A language is a way to specify a design and a set of constraints on that design. It may have one or more underlying MOC, and therefore differs from a MOC. *Denotational semantics* gives meaning of the language in terms of operations [124]. *Operational semantics* gives meaning of a language in terms of actions taken by some abstract machine (e.g. Turing machine). A *process* is an instance of a sequential machine, which can be implemented using hardware or software.

We use the definitions from the tagged system model [121] to describe the properties of a model of computation. A *value* is a piece of information (data) that is transmitted within the system. A *tag* is used to model time, precedence relationships, synchronization points and other properties associated with timing in the system. An *event e* is a value and a tag i.e.

given a set of values V and a set of tags T, e is a member of $T \times V$. A *signal* s is a set of events. A signal can be thought of as a member of the powerset of $T \times V$ or as a subset of $T \times V$. A set of signals is denoted by S, where $S =$ powerset($T \times V$). A tuple s of N signals is denoted by S^N. A process P is a subset of S^N. A particular signal s is a *behavior* of a process if $s \in P$. Therefore a process is a set of possible behaviors. We give graphical description of the relationships between these definitions in Figure 14.

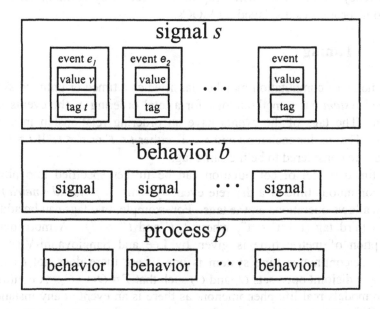

Figure 14 Venn diagrams of the notations in the tagged signal model.

There are two important properties for embedded system models of computation – the description of time and concurrency. Many embedded systems are concurrent in nature. Therefore, any MOC used to represent an embedded system must have concurrency as a first class citizen of the MOC. In addition, the representation of time is of extreme importance. An embedded system has a close relationship with its environment; hence the system often must be able to deal with physically occurring events. Yet, digital systems have traditionally used abstractions to remove time from the implementation of the system. This is most likely due to the programmer's sequential thought process and partly due to viewing the underlying system as a Turing Machine. Digital systems reduce the notion of time to finishing a sequential task as fast as possible; there is little thought of hard deadlines for computation. Yet, embedded systems face such strict computational deadlines. In fact, a correct answer is often of no significance in an embedded system unless it is delivered on time. An on-time answer that is partially correct is of more importance in an embedded system. In the Turing view of computation, any answer that is not correct is wrong. Even the underlying theory of MOCs for embedded systems is far different from the theory of traditional computing systems. We will first discuss the methods for timing a MOC and then talk about the representation of concurrency. Finally, we will wrap up the section by defining some other "random" properties displayed by MOCs.

2.1   Timing

A model of computation may be classified as a timed or untimed system. A *timed system* has a methodology for a total ordering of the events in the system. The tags of the events have an ordering relationship in a timed system. That is, for each tag t and $t \in T$, either $t < t'$ or $t' < t$. If $t = t'$, then the tags are considered to be the same tag.

A timed model of computation can be further specified into physical time, continuous time or a discrete event system. A *physical (metric) time* system allows operations on the tags. For example, two tags can be added to form a third tag ($t + t' = t''$, where t, t' and $t'' \in T$). A more detailed description of metric time is given by Lee and Sangiovanni-Vincentelli [121]. A continuous time system is one where there does not exist two nonempty disjoint open sets O_1 and O_2 such that $T = O_1 \cup O_2$. A continuous system models real life phenomenon, as there is an event at any instance of time. Differential equations are often used to model continuous systems.

A *discrete event* system has a timed tag system for all processes P. And for all $s \in P$, $T(s)$ is order isomorphic to a subset of the integers. Again, we refer the reader to [121] for more details. Essentially, a discrete event system has a countable number of events between any two tags. That is, given any two tags t and t', we can enumerate the every tag – in a specific defined order – that occurs between t and t'. More tersely stated, between any two finite time stamps, there are a finite number of time stamps. Most digital simulators use the notion of a discrete event system.

A *synchronous system* has a consistent notion of time throughout the entire system. If an event occurs somewhere in the system then that same event will occur everywhere at that time in the system. The event may be ignored, but it will occur. More formally, two events are synchronous if they share the same tag. Two signals s_1 and s_2 are synchronous if every event is s_1 (s_2) is synchronous with an event in s_2 (s_1). A process is synchronous with another process if every signal in the processes is synchronous.

An *untimed system* maintains only a partial ordering on the tags of the system. The tags in an untimed system form a partially ordered set. The ordering of the tags represents some kind of causality or synchronization point in the system. It is often hard to keep an overall notion of exact time across a system. This is especially true in distributed systems such as sensor networks. In this case, it is extremely expensive to coordinate the nodes of the network to a single clock. Even the synchronous nature of VLSI systems is reaching the point where it is becoming prohibitive to maintain a synchronous timed system; the chip geometries are becoming too small, in effect making the dies large enough that it takes several clock periods to span a chip. This is exacerbated by deep submicron effects like inductance and capacitance that further complicate the clock routing problem.

Many MOCs have the notion of an untimed system. For example, rendezvous models like communicating sequential processes (CSP) as well as process networks are untimed MOCs. We will discuss these models in detail in the following sections.

2.2 Concurrency

The second important aspect of embedded systems is the notion of concurrency or parallelism. Hardware is intrinsically parallel and software tends to lean towards parallel implementations in order to increase the performance of the applications that run on the system. Concurrency plays a major role in the description of an embedded system. And therefore, the concurrency allowed by the model of computation is a major factor in choosing a particular model to describe an application.

The fundamental limitation to concurrency is communication between two processes. A set of independent processes can be deemed "embarrassingly parallel" and a system can be easily constructed to efficiently execute any such set of processes. The problem with concurrency exists when there is an exchange of information between two (or more) or processes. We define *communication* as the exchange of information between processes. There are many possible ways to perform communication. Most often, we use some sort of storage element (e.g. register, bus, memory) to pass the data. The communication schemes differ in the mechanism used to access the storage data. We attempt to describe some of the prevalent methods of communication as used by models of computations for embedded systems. The method of timing often determines the type of communication in the system. Timing and communication – hence concurrency – are tightly interleaving in a model of computation.

Message passing is a particularly important method of communication whose roots lie in parallel computer architectures. Message passing can be divided into two separate categories – rendezvous and asynchronous message passing.

Rendezvous (sometimes called *synchronous message passing*) is an important method of communication in which a set of processes halt at a predefined point and exchange information. We will use the term rendezvous to describe this type of message passing because two processes that use rendezvous are not necessarily synchronous in the way that we defined synchronicity earlier. This is an example of different communities using the same words to have slightly different semantics. The processes run independently up until that point, exchange data at the rendezvous point and continue independently after the point. The processes communicate with instantaneous transactions; the read and write occurs simultaneously so there is no need to allocate any space for the data. If one process reaches the rendezvous point before the other process, it will stall until the other process is ready for communication. Rendezvous models are particularly well-matched to applications where resource sharing is a key element, such as client-server database models and multitasking or multiplexing of hardware resources. A key feature of rendezvous-based models is their ability to cleanly model nondeterministic interactions. [125]

Asynchronous message passing is broadly defined as communication between two processes that uses some sort of channel or storage element to pass the information. The main advantage of asynchronous message passing is that the sending process does not have to stall for the other process. The sending process places the information in a location that is mutually accessible and the receiving process can retrieve that information when it is ready to handle the data. Of course, the receiving process may have to stall if there is no data for it to consume, i.e. the sending process has not yet produced data.

There are many different ways to asynchronously transfer the data between two processes. We will discuss two such methods, first-in, first-out (FIFO) buffers and shared memory.

First-in first-out (FIFO) is a queuing discipline in which data in the queue leaves the queue in the same order in which it arrives. There are two types of FIFOs used in the embedded MOCs that we will describe. A *bounded FIFO* limits the amount of data that can be stored in the queue at any given time. An *unbounded FIFO* can store an unlimited amount of data. The actual realization of a FIFO in hardware can be accomplished in various manners. Most often, a buffer is used to implement a FIFO. The unbounded

FIFO is a nice abstraction that guarantees that no data will be lost. Clearly, the unbounded buffer size is not implementable in hardware.

Shared memory communication uses a common memory space to pass data between processes. A shared memory transaction can be of arbitrary size and can be as large as the processes' address space. Unlike a FIFO, it allows multiple nondestructive reads to occur to the data. Therefore, a process can communicate the same information to multiple other processes by performing only one write. A write will delete any previous data that was written to the same address. The writing of data to a shared memory space must be organized so that no two processes write to the same address at the same time. An arbitration scheme such as a semaphore must be implemented with any shared memory component.

2.3 Other Properties

Computational models can also be distinguished by other properties such as their static and dynamic nature, determinism and reactivity.

A model of computation that is *static* if it cannot create processes or additional hardware resources at runtime. Static models are advantageous because we can do heavy analysis and optimization at compile time. A static synchronous language can be represented in a finite number of states, making verification easier. A *dynamic* MOC allows unbounded number of resources at runtime. For example, any model that assumes an unbounded FIFO is dynamic. Normally, a dynamic MOC is considered unrealistic for a system, though reconfigurable systems are changing that notion. Due to the nature of the reconfigurable system, it can dynamically create and remove resources at runtime. The SCORE model of computation, which targets reconfigurable systems, is dynamic and allows for reconfiguration to handle the dynamic nature of the model. We will further discuss the SCORE model in a later section.

Determinism is the ability to reason about the outputs of a MOC based on its inputs. In a deterministic MOC, the inputs to the system fully specify the system's outputs. More specifically, a process is determinate if every behavior of the process can be achieved by applying some input to the process. A non-deterministic MOC may seem like a negative quality, but on the contrary, it allows programs that can respond to unpredictable sequences of input events (e.g. embedded systems interacting with their environment) or can defer complete specification of the system until some later time.

Reactive systems are those that react continuously to their environment at the speed of the environment. Harel and Pnueli [126] and Berry [127] contrast them with *interactive systems*, which react with the environment at their own speed, and *transformational systems*, which simply take a body of input data and transform it into a body of output data [128]. A reactive system is often needed for embedded systems, particularly real time systems. Ideally a reactive system will maintain a permanent reaction with the environment i.e. it will never terminate and execute quickly with regard to some stimulus. Often an embedded system will lie somewhere between a reactive system and an interactive system. The system will have hard timing constraints in which it must respond. It is interesting to note that a "correct" reactive system is fundamentally the opposite of a correct Turing system. A correct Turing program is one that will ultimately halt; a correct reactive program will never halt. Determining whether a program will halt and correspondingly whether a program will never terminate is an undecidable problem [129].

3.   CLASSIC MODELS OF COMPUTATION

The importance of a specific property for embedded and reconfigurable systems depends on the application(s) that one wishes to implement on that system. If an application is control dominated, we will use a MOC that efficiently models control flow. On the other hand, an application that has a large amount of data flow will use a MOC that can handle such a property. In this section, we will describe several fundamental models of computation. These MOCs have nice properties geared towards a specific type of application. Therefore, one of these MOCs can be directly used to model an application if that application exhibits the same properties as the MOC.

More recently, researchers have realized the importance of hybrid models of computations. Hybrid MOCs combine two or more fundamental MOCs in an attempt to benefit from the good properties of both of the MOCs [125]. We will discuss some of these hybrid models in the following sections. In this section, we will only focus on the classic models that are relevant to embedded and reconfigurable system design.

Concurrent sequential processes (CSP) were introduced by Hoare [130] in an attempt to make input, output and concurrency first class citizens of a programming language. He created a succinct notation for adding these three concepts into a programming language. CSP uses rendezvous for structuring the communication between processes. Two or more sequential processes proceed autonomously, but at certain points in their control flow, they coordinate so that they are simultaneously at specified points.

Communication occurs between two processes A and B when process A specifies process B as an output and process B specifies process A as an input. There is no buffering of data. A process that outputs data to another process must wait until the input process is ready for the data. Processes are explicitly stated to execute in parallel. A parallel processes waits until every process in the set has finished executing before continuing. Processes can have guarded execution i.e. they can be predicated by a guard expression and will execute only if that guard expression holds. A negative property of CSP is that it allows for the possibility of deadlock.

Discrete event (DE) models have a countable number of tags. The time stamps that appear in the system can be enumerated in chronological order. Every event is placed precisely on a globally consistent time line [125]. Discrete event models are timed systems that maintain the property that any two events have a discrete number of events between them. In other words, during the time between any two events there are a quantifiable number of other events.

A DE system uses events to communicate values within the system. The events occur at a distinct time. Since there is a globally consistent concept of time throughout the system, the values of the events may be used to denote a piece of data that some component in the system wishes to exchange. The global notion of time does not necessarily imply the use of a global clock. Furthermore, there is no guarantee that every event will be seen by the receiver because the receiver may not be sensitive to that event when it is emitted.

Discrete event models are excellent descriptions of concurrent hardware, although increasingly the globally consistent notion of time is problematic. In particular, it over-specifies (or over-models) systems where maintaining such a globally consistent notion is difficult. This includes large VLSI chips with high clock rates.

Cycle driven models are a subset of DE models. Cycle driven models associate a clock with each component. Computations are assumed to occur only at a specific time of the clock cycle, e.g. the rising edge of the clock. Therefore, a cycle driven models is a DE model with the restriction that events occur at regular intervals.

A *synchronous/reactive (SR)* model of computation describes a system as a set of concurrently-executing synchronized modules. These modules communicate through signals that are either present or absent in each clock tick. The presence of a signal is called an event. The value of the event carries a piece of data such as an integer. The modules are reactive in the sense that they only perform computation and produce output events in instants with at least one input event. A synchronous/reactive model of computation uses a global clock to define distinct points when a signal may have a value. It behaves much like discrete event model with the exception that a signal may not have a value at every clock tick. Examples of SR language include Esterel [131], Signal [132], Lustre [133], and Argos [134].

SR models are excellent for applications with concurrent and complex control logic. Because of the tight synchronization, safety-critical real-time applications are a good match. However, also because of the tight synchronization, some applications are overspecified in the SR model, limiting the implementation alternatives. Moreover, in most realizations, modularity is compromised by the need to seek a global fixed point at each clock tick [128].

A *process network (PN)* is an asynchronous message passing model where input sequences are mapped to outputs by processes. Kahn [135] developed and gave many of the formal underpinnings to the process network, hence PN are sometimes referred to as Kahn process networks.

The communication in a process network is done through a channel. A channel is a possibly infinite sequence of data objects or tokens. The channel can be viewed as an unbounded, unidirectional FIFO buffer. The writes to the channel are non-blocking, i.e. they always succeed, while the reads from a channel are blocking. In other words, a process must wait for an event to be placed in the channel before it can complete the read to that channel. Each token is written to a channel exactly once and subsequently read exactly once. The state of the process network is determined by the number of tokens in the channels.

The processes of the process network are concurrent. Processes are a continuous mapping from input streams to output streams. Kahn process networks are important models that guarantee lossless communication at the highest level of abstraction by assuming an ideal buffering scheme that has unbounded buffer size. Clearly, the unbounded buffer size is a "nonimplementable" way of guaranteeing losslessness [10].

Dataflow process networks are a model of computation used in many digital signal processing (DSP) applications. Dataflow process networks are known to be a special case of Kahn process networks [136]. They are an untimed model; they only have a partial ordering of the operations in the system.

The dataflow process network is specified by a directed graph where the nodes (called *actors*) represent computations and the edges represent totally ordered sequences of events (*streams*). The actors take a set of input tokens, perform some sort of computation on them and produce output token(s). This action is called a *firing* of the actor. A firing can be viewed as an indivisible quantum of computation. It is possible that the firing of an actor may consume many input tokens from a particular stream or channel. Additionally, the firing may depend on a set of conditions called the firing rules of the actor. For example, the actor may not fire unless there exist at least one input token on all of its input channels.

Dataflow process networks guarantees the safe arrival of every token and that the sequence of tokens on each communication path is defined. However, there is no global ordering for tokens in unrelated design. The process of determining a ordering of the firing is an important problem in dataflow process networks. The firing scheduling problem should avoid ordering that cause deadlock in the process network. Another desirable property of a schedule is a finite number of tokens in every channel over the entire schedule i.e. the number of tokens in the channels should be bounded. An even more refined version of dataflow process networks allow for guaranteed static scheduling of the firings. This particular subset of dataflow is called synchronous dataflow.

Processes in a *synchronous dataflow* [137] model consume and produce a fixed number of tokens on each firing. Synchronous dataflow is not synchronous in the manner which we defined synchronicity. In fact, the authors have even admitted to this misnomer stating that static dataflow would probably be a better description for this model of computation. Nevertheless, hindsight is 20/20 and synchronous data flow retains its historical name.

Synchronous data flow enforces strict firing rules; each input stream of a firing process must have at least one input token in order to fire. Synchronous data flow has the property that a finite static schedule can always be found to bring the data flow graph back to its original state i.e. there exists an *admissible* (actors fire only when firing rules are valid) and *periodic schedule* (the network returns to its initial state while firing each actor at least once) for every instance of synchronous data flow. Additionally, it is possible to statically determine the boundedness of a particular schedule. However, even different admissible, periodic, bounded schedules will produce variances in the code size, buffer size and pipeline utilization of the schedule.

4. EMBEDDED AND RECONFIGURABLE MODELS OF COMPUTATION

4.1 Control Data Flow Graph

The model of computation used describe an application greatly affects process of system synthesis. The more a model exposes the underlying hardware, the better the application programmer can customize the application to the system. Yet, the additional semantics of the model can limit the productivity of the programmer, as they must learn exactly what the semantics mean to the implementation. In addition, the semantics may enable the programmer to limit the design space; a design specified with less restrictive semantics may result in better system performance if the system synthesis engine is highly optimized.

A data flow graph $G_{dfg}(V,E)$ is a directed graph that represents a set of operations O and their dependencies. For each operation $o_i \in O$ there is a vertex $v_i \in V$ i.e. there is a one to one correspondence between the operations and the vertices. Each operation has set of input operands and produces a set of output operands. If an output operand of operation o_i (vertex v_i) is in the set of input operands of operation o_j (vertex v_j), there is a directed edge $e(i,j)$. The behavioral description has arithmetic and logical operations. The data flow graph is an instance of a dataflow process network. Furthermore, it can be viewed as a restricted synchronous data flow model. The vertices are actors and the firing rules are limited to the case where there is exactly one token for each input stream and the actor produces exactly one output token.

A *control data flow graph (CDFG)* is a type of the data flow graph that contains information related to branching (if-else constructs) and repetition (while loops). There are many different models for CDFGs [16, 138, 139]. We choose to represent the CDFG as a two-level hierarchical sequence graph. A sequence graph is a hierarchy of data flow graphs [16]. The highest level in the hierarchy represents the flow of control in the application. The vertices of the highest level are associated with a data flow graph. The edges correspond to a possible flow of control in the application. Consider Figure 15, an if-then-else construct. The top level CDFG of the hierarchy has four vertices. Each vertex contains a data flow graph that is executed when the control flow reaches that vertex. The control edges model the flow of control of the if-then-else construct.

The CDFG offers several advantages over other models of computation. Most compilers have an internal representation (IR) that can easily be transformed into a CDFG. Therefore, this allows us to use the back-end of a compiler to generate code for a variety of processors. Furthermore, the techniques of data flow analysis (e.g. reaching definitions, live variables, constant propagation, etc.) can be applied directly to CDFGs. Finally, many high-level programming languages (Fortran, C/C++) can be compiled into CDFGs with slight modifications to pre-existing compilers; a pass converting a typical high-level IR into control flow graphs and subsequently CDFGs is possible with minimal modification. Most importantly, we believe that the CDFG can be mapped to a variety of different microarchitectures. All of these reasons make CDFGs a good MOC for investigating the performance of mapping different parts of the application across a wide variety of SOC components.

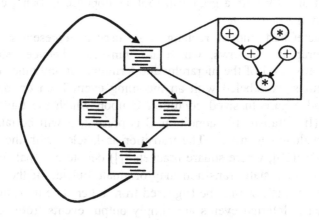

Figure 15 A Control Data Flow Graph

On the other hand, CDFGs only have the ability to describe instruction level parallelism. In order to specify a higher level of parallelism, another MOC must be used. But, we could embed CDFG into another MOC – one that can describe a higher level of parallelism. For example, we could embed CDFGs into finite state machines (FSM). Edward Lee's (UC Berkeley) *charts [140] do something similar; they embed SDF graphs into FSM.

4.2   StateCharts

David Harel's *StateCharts* model [141] was developed to overcome some of the difficulties of the finite state machine model for embedded systems. Specifically, the FSM model does not allow for hierarchical abstraction of states, which is unwieldy. Additionally, concurrency in execution is not well handled by finite state machines, which are inherently sequential by design. In order to represent multiple concurrent states (e.g. states C and F occurring simultaneously), an FSM must merge the concurrent states into a single state (e.g state CF). The potential simultaneous states are a subset of a power-set of the original set of states. Thus, parallelism at the FSM level may lead to an exponential explosion of states.

The StateCharts model of computation is an extension of the finite state machine model. StateCharts are hierarchical, which enables modular design of complex systems. Additionally, it contains constructs to enable easy representation of concurrency and broadcast communication, which are often lacking in other hierarchical state machine models. The main goal of StateCharts was to be a human-readable design model, which could be manipulated directly with a graphical tool, as opposed to being completely programmed in ASCII.

In the StateCharts model, rectangles are used to represent states at any level of hierarchy. Arrows, which are transitions between states, may originate at any level of the hierarchy and terminate at any other level. All state transitions are labeled with an incoming event I, an optional output event O, and a parenthesized condition C upon which the transition will predicate. (If a transition's condition C is present, it will be taken if and only if C evaluates to true.) The transition is labeled with the following string: I [/ O] [(C)], where square brackets ([]) denote optional fields. The events that drive a state transition may originate outside of the state-chart (i.e. external events), or may be triggered from other state transitions inside the state-chart. Internal events are simply output events from some other state. This side-effect mechanism is the method by which states communicate, and is a form of instantaneous broadcast communication. One state may generate an output event, and all other states simultaneously sense this event and may react to it. Although this ability to model internal and external events is very powerful, it may lead to some potential design complications. For instance, if an external event triggers an internal event which affects other states, a large chain reaction of side effects may be followed, leading to a new set of states. Since these chain reactions take zero time to occur in the model (as all communication is instantaneous), a complex set of semantics was produced by Harel et al [142] to resolve chains in a deterministic fashion. A discussion of these semantics is beyond the discussion in this book.

States may be hierarchically combined via XOR decompositions and AND decompositions. These decompositions are visually represented as rectangular states drawn around other states (see Figure 16). An XOR decomposition of two states s_1 and s_2 means: "the machine is either in s_1 or in s_2, but not both." Similarly, an AND decomposition of two states s_1 and s_2 means: "the machine is simultaneously in s_1 and in s_2." AND decompositions, like XOR decompositions, may describe the decomposition of any number of states (although for simplicity we have just described decompositions of two states), and they may be used on any hierarchical level. Additionally, within an AND decomposition or an XOR decomposition, entire sets of states may be placed (as opposed to a single state). A start state (or set of start states) is represented by an arrow to the start state (as in the FSM model) as well as other optional arrows to any states that hierarchically contain the start state.

XOR decompositions are typically used to economize arrows in the state diagram. When transitions from two or more states go to the same destination state on the same input event, these states may be clustered into a new state, and the arrow replaced by a single arrow. This replacement of multiple transitions by a single transition maintains visual simplicity and modularity within the StateCharts model.

AND decompositions are used to represent concurrency within the StateCharts model. States clustered via this method are said to execute simultaneously. This avoids the FSM model's exponential state explosion mentioned earlier.

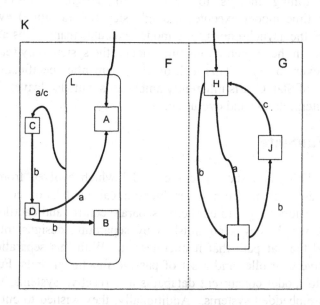

Figure 16 An Example StateCharts Program

A small StateCharts program is represented in Figure 16. In this program, K is the top-level state. Within K, an AND decomposition is used to execute states F and G in parallel. F contains sub-states C, D, and L (which itself contains states A and B). G contains substates H, I, and J. In state L, an XOR decomposition is used to group states A and B together. That is, from either state A or B, if event a is received, a transition is taken to state C and output event c is produced. When this transition is taken, every component in the StateCharts program receives event c simultaneously. Thus, if concurrently executing state G is in sub-state J as the transition emitting event c is taken, a state transition will occur from J to H within G. The program starts in states A and H simultaneously.

Although the StateCharts model is a very powerful hierarchical and concurrent extension to finite state machines, its broadcast event communication leads to a potentially high amount of inter-state dependency. This may result in some difficulty resolving the next state(s) of the machine.

As mentioned earlier, a set of complex rules exist whereby the next state may be resolved, but it is an unwieldy task for a designer to weave through these dependencies in highly complex programs. Furthermore, as some of the semantics of StateCharts are left open, many different implementations of StateCharts now exist, each with some unique set of semantics. For instance, STATEMATE [143] an implementation that Harel took part in, allows two timing models to be used in design. STATEMATE's synchronous time model executes a single step (or transition) every time unit, whereas the asynchronous time model executes many steps at a single point in time. In the asynchronous time model, the system only reacts when an external event occurs. This lack of exact semantic specification leaves some aspects of StateCharts inherently ambiguous, but also leaves the model flexible to interpretation and evolution.

4.3   FunState

FunState [144] is a state machine model which evolved from Harel's StateCharts as well as from other hierarchical models of computation. FunState's creators sought to explicitly separate control and dataflow within their model, which they accomplished by separating designs into a state machine and the datapath that it operates on. With this separation into a purely reactive controller and a set of passive functional units, FunState is easily able to model concurrent datapaths and reactive systems, which are prevalent in embedded systems. Additionally, they wished to cut all noise out of their model, and only represent design characteristics relevant to design methods such as scheduling and verification. This choice, while detracting from the model's completeness, adds to its simplicity and usefulness in embedded system codesign.

The basic FunState Model consists of a network N (representing a datapath) and its controlling finite state machine M. N is defined by three sets F, S, and E ($N = (F, S, E)$). F is a set of functions which are uniquely named. Functions represent computations which consume some number of inputs and produce outputs. S is a set of storage elements, which can be either FIFO queues of unbounded length or registers (a mapping from an address to a value). E is a set of directed edges, where each edge is either from a storage unit to a function, or from a function to a storage unit. Therefore, storage units and functions form a bipartite graph.

Data is represented as a set of tokens, which are moved between functions and storage elements. Similar to the dataflow process model described above, every function is an actor which operates on tokens from storage when it fires, and produces output tokens to be stored upon completion. The edges of the network act as streams moving data from the actors to/from storage elements.

The finite state machine M is a synchronous reactive machine similar to Harel's StateCharts model. State transitions are arrows labeled with predicates and actions. Predicates are a set of Boolean conditions which must be true in order for the transition to occur. For instance, a transition may be predicated upon a certain number of tokens in a queue, or a value in a register. The actions are a set of functions in N that will fire upon transition. Thus, M controls the sequence of functions executed within N.

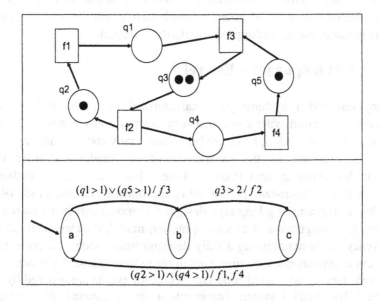

Figure 17 FunState network (top) and controlling state machine (bottom)

The state machine begins execution at an initial state, and all storage elements in N are filled with their initially valued tokens. All predicates of transitions out of the current state are evaluated. If no transition's predicate evaluates to true, execution stops. Otherwise, one non-deterministically chosen state transition whose predicate condition is true shall be taken. All functions in the action set of the transition are fired in a non-deterministic order. Every fired function removes tokens from its input storage units and produces tokens for its output storage unit. Execution proceeds as just described until the machine enters a state in which execution stops.

A FunState network and finite state machine are shown in the top and bottom (respectively) of Figure 17. The network consists of four functions (*f1* through *f4*) and five storage elements (*q1* through *q5*). Each edge of the network connects a function (drawn as a rectangle) to a storage element (drawn as a circle). The black dots inside certain storage elements indicate data tokens stored there. For instance, $q3$ contains two data tokens and $q2$ and $q5$ each contain one data token. The finite state machine on the bottom of Figure 17 has three states: a, b, and c, which control execution of functions in the network. For instance, if the state machine is currently in state c, and $q2$ and $q4$ each have more than one data token, then functions $f1$ and $f4$ are executed simultaneously.

The state machine M may be designed hierarchically, with AND and XOR decompositions similar to those of StateCharts. Additionally, the network N may contain sub-FunState models (a sub-network and a sub-FSM) embedded as function nodes. This hierarchy can be made arbitrarily deep, as in StateCharts, leading to modularity in design.

4.4   S/R languages – Esterel

Many embedded systems are characterized by permanent or semi-permanent interaction with a continuously changing environment in which they are embedded. Often, a system built from many concurrently executing modules is required, as the parallelism of the modules enables faster execution for better system response-time. For this class of embedded systems, a synchronous-reactive model is appropriate. *Esterel* [145, 146] is a high-level programming language designed to model synchronous-reactive systems. As a language and as a computation model, it successfully models concurrency while maintaining a fully deterministic system. Every event in the system is synchronous, as time tags in the system are totally ordered, and every event has an associated time tag. Thus, all events are ordered by their time tag. The Esterel system represents a set of concurrently executing processes, which communicate with each other by sending signals. Additionally, the system receives signals from its external environment. At an instance in time, a signal is said to be either present or absent.

The Esterel language is composed of *signals* $S_1...S_n$, and *events* $E_1...E_m$. Each event E_j is a set of signals. A signal which belongs to an event is said to be present in that event, otherwise it is absent. Execution of an Esterel program derives a sequence of output events from a given sequence of input events. The production of an output event as a result of an input event is called a *reaction*. The flow of time in Esterel is defined by the sequence of reactions which occur during execution.

In Esterel, a statement takes time to execute if and only if it specifies this. These statements are known as *temporal statements*. Temporal statements take exactly as much time as they specify. They include *triggers*, which specify a set of statements that execute if and when a specific event happens. Temporal statements also include *watchdog* statements, which specify a set of statements to execute until either the statements are finished or until a specific event happens. In this manner, triggers are a pretest for the occurrence of an event, and watchdogs are a post-test. As an event could be a signal from a clock or a timer, it is possible to create any kind of wait or timed loop in this fashion.

Statements which are not temporal execute in zero time. In other words, these computations are instantaneous. Instantaneous computations in the Esterel system include memory updates, communication between processes, and conditional structures. Assignment statements, signal emissions, loops, selection statements, traps, and exits are all instantaneous language constructs. All internal communication in an Esterel system is instantaneous, and is broadcast between senders and receivers. This implies Esterel's *Global Coherence Law*: a signal is said to be present at a given instant if and only if it is received from the environment or if it is internally communicated at that instant.

Concurrent execution is modeled in Esterel via a parallel statement. For instance, the expression "s1 || s2" means: "begin executing statements s1 and s2 in parallel." Parallel statements are started simultaneously, and the Esterel program waits for all concurrent statements to finish execution before continuing, unless one or more concurrent statements exits via a trapped exception.

Despite its support of concurrency, Esterel's communication semantics provide fully deterministic programs. Esterel is a highly versatile model of computation, as it can be easily translated to automata, sequential software, or boolean circuitry [147]. It is a natural model for reactive embedded processes, as it supports temporally parameterized programs. Although Esterel relies on the programmer to specify the exact parallelism in an algorithm, Esterel (like other hardware description languages) could be envisioned as the target of a higher-level compiler, which converts sequential algorithmic code in C (for instance) into a parallelized Esterel specification.

A small piece of Esterel code is shown in Figure 18. The code declares a module named EX. EX has two input signals (ON and CLK) and two output signals (N1 and N2). EX also has a single integer variable named COUNT, which is initialized to 0 in its declaration. At the beginning of execution, signals N1 and N2 are emitted, each with a value of 0. Then execution is halted until the signal ON is received. At this point a loop is entered, which executes every time an event happens on signal CLK. The body of the loop is a parallel statement, with two blocks of code (encapsulated in []) executing simultaneously. The first block increments the variable COUNT and waits for signal N2 to be emitted. Concurrently, signal N1 is emitted with the current value of variable COUNT and signal N2 is emitted with the

current value of COUNT plus constant 1. The module effectively counts the
number of times a clock signal occurs.

```
module EX:
input ON (integer), CLK (integer);
output N1 (integer), N2 (integer);
var COUNT :=0 : integer in
   emit N1(0);
   emit N2(0);
   await ON;
   loop
       [COUNT := COUNT + 1;
        await N2]
    ||
       [emit N1(COUNT);
        emit N2(COUNT + 1)]
   each CLK;
end var
end module
```

Figure 18 A Sample Esterel Program

4.5   SpecCharts – Program State Machines

Although VHDL and other hardware description languages are
frequently employed in the design of embedded systems, their programming
models typically have difficulty representing certain desired attributes of
these designs. Specifically, VHDL is often awkward in capturing concurrent
behavior (forking is not a supported operation, for instance) or unstructured
jumps, useful in state transitions for state machines and in exception
handling. The *SpecCharts* language [148] (which was later redesigned as
the SpecC language) added these abilities to VHDL in the form of additional
programming constructs. The model of computation used by the SpecCharts
language was the *Program State Machine (PSM)* model.

The PSM model is a hierarchical and concurrent Finite State Machine
(FSM) model, where a leaf state (one not composed of other states) can be
described as a program. No limitations exist on the program's size or
complexity. Thus, the PSM model subsumes the FSM model as well as a
programming language model. Formally, a *PSM* is a pair $<I, P_{root}>$, in which
I is the set of all input/output ports, and P_{root} is the root program state. A
program state P is defined as a triplet *<decls, status, comp>*, where *decls* is a
set of variable and procedure declarations inside P, status is the state's

current execution status (either *inactive*, *executing*, or *complete*). *Comp* is either a leaf program state (a program), a concurrently-composed program state, or a sequentially-composed program state. A concurrently-composed program state is a collection of program sub-states, in which all program sub-states execute concurrently. Likewise, a sequentially-composed program state is a collection of program sub-states which execute in sequential order. Between sequentially executing sub-states are transition arcs, which are analogous to control-flow edges, as they dictate which program state shall execute next based on a condition. The if-then-else construct pictured in Figure 15 could be represented in PSMs as two transition arcs, one which transitions to the left path, and the other one (with the opposite condition) transitioning to the right path. A transition arc between two states (first and second) is either transition-on-completion (TOC) or transfer immediately (TI). A TOC arc means that the second sub-program state shall execute when the first state has completed, as long as the arc condition is satisfied. A TI arc means that whenever the arc condition is satisfied, the second state begins executing (and the first state ceases activity, even if it has not finished executing).

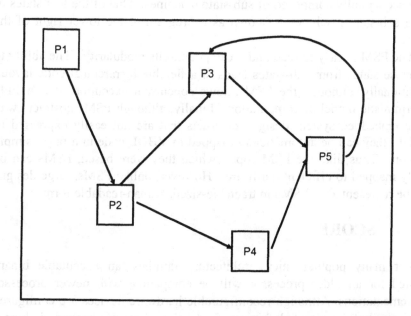

Figure 19 A Program State Machine (PSM)

When a PSM begins executing, its P_{root} status is set to *executing*. Whenever a program state begins executing, it will immediately exhibit appropriate behavior. For instance, when a concurrently-composed state is executed, all of its sub-states are simultaneously executed. When a sequentially-composed state is executed, its first sub-state is executed. When a leaf state is executed, its statements are executed. As soon as a leaf-state's last statement is executed, it is said to complete. Likewise, a concurrently-composed state completes when all of its sub-states have completed, and a sequentially-composed state completes when an arc-transition occurs to a special *complete* sub-state. If a program state is inside a sequentially-composed state, its completion will cause a transition to a new state to occur if a transition arc from the current state to the new state has its *cond* met. If more than one state transition can occur, the one closest to the front of the transition arc list is traversed. The PSM model supports exceptions as activated TI arc transitions, as a TI is an immediate jump out of the current state into some other (without finishing the execution of the current state).

Figure 19 shows a sample PSM. The start state is clearly labeled with an incoming arrow which has no originating state, as in the FSM model. The second state (the one immediately reachable from the start state) is expanded to show that it is a concurrently composed program state, with each parallel state sequentially composed of sub-state machines. One of the leaf states of these sub-state machines is shown as a program in the lower right of the figure.

The PSM is easy to read and create, due to its modularity. The ability to compose states from sub-states leads to a flexible hierarchical code layout. Additionally, although the PSM allows concurrent execution, it is a fully deterministic model of computation. Finally, although PSMs abstract away some embedded system design constructs that are not easily expressed in VHDL, they can be automatically mapped to VHDL code in a fairly simple fashion. Thus, like the FSMs upon which they were based, PSMs can be easily mapped to efficient hardware. However, unlike FSMs, large designs can be represented as PSMs in tractable-sized, human-readable form.

4.6   SCORE

For many popular micro-architecture families, an executable binary created for an older processor will be compatible with newer processor implementations. Although reconfigurable hardware devices are configured with similarly generated binary files, this sort of backward binary compatibility has not been accomplished. This difficulty of porting designs between device generations has been a significant obstacle to the widespread adoption of reconfigurable devices. *SCORE* [149] is a reconfigurable computation and execution model designed to overcome these obstacles. The SCORE computation model is an extended variation on the Kuhn process network model, with a fully dynamic set of processes and streams, and logically unbounded communication channels dynamically mapped onto physical resources. The SCORE execution model implements this dynamic behavior via a runtime environment, which can reconfigure the physical hardware device. This makes the SCORE computation model one of the most flexible in this study.

SCORE's computation model may be seen as a graph. Each node corresponds to either a computation operator or a memory block, and each edge represents a data stream between two nodes. A SCORE computation operator represents any algorithmic transformation which translates input data to output data. This algorithm could be, for example, a multiplier, an adder, a multiply-accumulate operator, or an FIR filter. They may be built as behaviorally primitive operators or compositions of these operators. For instance, an FFT operator may be composed of other operators, among them an adder operator.

Computation operators come in two flavors: Finite State Machine (FSM) nodes and Turing Complete (or TM) nodes. An FSM node represents each state's inputs as a set to be read from input data streams. Once all inputs are present, the FSM executes, transitioning to a new state and optionally emitting output. Additionally, sets of input and output streams may be closed during execution. FSM nodes also have a distinguished done state which, if entered, will signal the FSM's completion and allows the FSM to remove itself from further computation. Put differently, an FSM's done state effectively removes the FSM from the computation graph. A TM node is a special case of FSM node that has the additional ability to create new nodes (FSM, TM and memory blocks) and stream edges in the graph. In other words, TM nodes represent computation that can allocate storage and also allocate new computation.

A SCORE operator's hardware size is implementation dependent, and is not specifically limited in the SCORE model of computation. SCORE operators are implemented as compute pages (CPs) in the execution framework. A SCORE compute page is a fixed-size block of physical reconfigurable logic, such that a single compute page represents a single schedulable unit of execution. Each SCORE computation represents a set of active compute pages. Each compute page may encompass the implementation of one or more computation operators, and a single computation operator may be implemented as one or more compute pages. (The partitioning and mapping of SCORE computation operations onto physical compute pages is a duty of the SCORE compiler.) As a single compute page is the smallest SCORE-allocatable piece of physical reconfigurable logic, the physical size of a compute page is largely dependent on the target hardware. For instance, on an FPGA device, a compute page may be a logical grouping of a set of 4-LUTs. If the full set of compute pages in the SCORE program does not fit on the reconfigurable device, then the compute page configurations are time-multiplexed onto the reconfigurable fabric. The dynamic configuration of compute pages onto their corresponding logic allows the physical device to execute a program of unbounded size.

A SCORE memory block corresponds to a contiguous allocation of memory segments on the physical hardware. A memory segment is a fixed-size contiguous amount of memory, the unit of storage size for data managed by SCORE's runtime environment. A memory segment's size may vary across physical hardware implementations. A single memory segment may be represent any physical amount of memory up to an architecturally defined maximum.

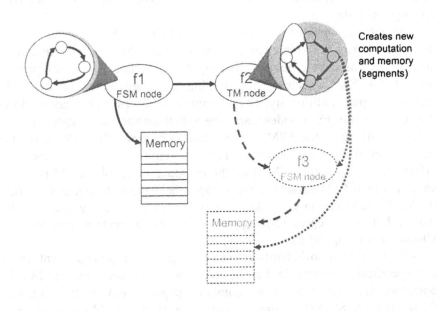

Figure 20 A SCORE Computation

A system of memory ownership exists in SCORE, whereby operators are given mutually exclusive ownership of memory segments. Each memory segment is owned by a single operator at a time. Operators may only access memory segments that they own. As TM operators are allowed to dynamically create new FSMs and TMs, they are also allowed to create new memory segments. When a TM creates a segment of memory, it passes it (i.e. transfers its ownership) to an FSM or TM that it previously created. Upon entering the done state, a TM or FSM disowns the memory segments which were passed to it, and the memory's ownership is returned to the creating TM. If an operator attempts to access memory it does not currently own, the operator will block until it regains ownership of that memory. At execution time, memory segments are mapped onto configurable memory blocks (CMBs), which are analogous to physical compute pages. Along with a set of memory segments, each CMB may also hold compute page configurations, state, and stream buffers.

Streams are channels of data which link data producer operators to data consumer operators or to memory segments. In other words, streams are the abstraction of data communication in the SCORE computation model, and they define where data will be logically routed. Each stream is logically represented as an unbounded queue of data tokens. To support this representation on actual reconfigurable hardware, streams are implemented as spatial connections until they become full (i.e. when a consumer consumes data slower than a producer sends data). At this point, a full stream is mapped onto FIFO buffer segments (located in CMBs), and the connection between operators is rerouted through this segment. FIFO buffer segments are also useful for communication between time-multiplexed compute pages. If both pages are currently available on the chip, the stream may be a spatial connection. Otherwise, if either a producer or consumer is not resident, a stream is sourced-from/sinked-into a FIFO buffer segment in some CMB.

A sample SCORE program is shown in Figure 20. At the start of execution, there are only two functions in the program: *f1* and *f2*. There is also one memory block, which function *f1* streams into. As *f1* is a simple FSM node, it executes its finite state machine (shown in the callout in the upper left of the figure), and then streams the output of its computation to function *f2*, which is a TM node. Then, *f2* begins execution of its internal FSM (shown in the upper right of the figure). One of the states inside *f2* dynamically creates a new function *f3* and a new memory block (pictured at the bottom of the figure). When *f2* finishes execution, it streams its output data to the new function *f3*, which in turn executes and streams its output into the new memory block.

SCORE enjoys many advantages, as it is a fully flexible model for any reconfigurable device. That is, the SCORE model of any program may fit onto any target device. (This is similar to the intermediate representation of a retargetable compiler, and its ability to be mapped into code for many target architectures.) When the SCORE executable model can fit onto a device, it enjoys a fully spatial architecture, which enables efficient, pipelined chaining. Although the program is certainly slower when configured onto a smaller device (on which all of it cannot fit at once), it is still executable on this device, which allows backward binary compatibility. The dynamic instantiation of memory and operators makes SCORE a very flexible model in terms of expressiveness. Additionally the compute-page and CMB abstractions allow coarse-grained hierarchical resource management, as each resource unit can be designed, placed, and routed individually offline, and then combined online. However, a tradeoff can be seen between SCORE's complexity and its completeness. The model places a heavy burden on runtime OS support on a sequential co-processor. The OS must schedule virtual pages and streams, assign pages to reconfigurable hardware blocks (and migrate them, when necessary), and perform routing. An online approach to these problems is required, as their implementation relies on dynamic field reconfiguration.

4.7   Streams-C

Streams-C [150] is an extension of Hoare's CSP model [130], which emerged from many years of compilation and synthesis work under Maya Gokhale at Los Alamos National Laboratory. It is a programming language of annotations and libraries callable from standard C code, which models stream-oriented FPGA applications. These applications tend to be characterized by fixed-size data, computationally intensive processes, and high data-rate. Synchronization of streams is an expected requirement, but it is assumed that this is only occasionally necessary.

With Streams-C, a programmer may declare processes, streams, and signals. Processes are independently executing objects, defined by a subroutine written in C. Each process has its own address space, and processes communicate via message passing. Much of the time, this message passing occurs via data streams, which are the only input/output parameters of a process's subroutine implementation. In Figure 21, a basic Streams-C program is represented as a directed graph, with vertices P1 through P5 representing processes, and each edge representing a stream from one process to another.

A stream represents data moving between processes, and is defined by its payload size and the semantics of its operation. Within a process, library functions may be called on streams in order to open, close, read, and write them. Also, a process may check a stream to see if it is empty.

In addition to streams, processes may communicate via the use of shared signals. These signals are typically used for synchronization of processes, in a similar manner to the use of signals in other models (e.g. Esterel). A process may include the statement **sc_wait**(<signal_list>) in order to wait on a list of signals. Upon reaching this statement, the process blocks until any one of the signals in the signal list is posted (i.e. activated). Processes may post signals using the **sc_post**(<signal_name>, <value>) command, which posts the identified signal along with an optional value.

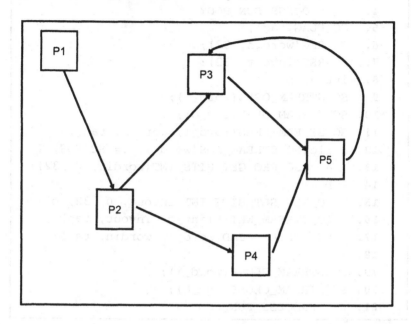

Figure 21 The Streams-C model: Processes interconnected by streams

In Figure 22, a sample Streams-C process is shown, which implements the functionality of a simple 32-bit inverter. The process has two streams, the input stream word_i and the output stream inv_o (identified as process streams by lines 2 and 3, respectively). These streams are accompanied by a one-bit flag, which is declared in the SC_FLAG statement on line 5. Each stream is attached to a 32-bit process port. The process ports wordin (connected to stream word_i) and invout (connected to inv_o) are each declared by an SC_REG statement on lines 6 and 7. Much of the body of the process is composed of stream operations, as word_i is read through the wordin port, assigned to integer a, and then inverted and assigned to integer b. The contents of b are then written to the outgoing stream inv_o via port invout. This process is continued as long as the end of the input stream is not reached (as tested on line 12). Before the process concludes, both streams are closed.

```
1.  /// PROCESS_FUN inverter
2.  /// INPUT word_i
3.  /// OUTPUT inv_o
4.  /// PROCESS_FUN_BODY
5.     SC_FLAG(tag);
6.     SC_REG(wordin, 32);
7.     SC_REG(invout, 32);
8.     int a, b;
9.     SC_STREAM_OPEN(word_i);
10.    SC_STREAM_OPEN(inv_o);
11.    SC_STREAM_READ(word_i, wordin, tag);
12.    while(SC_STREAM_EOS(word_i) != SC_EOS) {
13.       a = SC_REG_GET_BITS_INT(wordin, 0, 32);
14.       b = ~a;
15.       SC_REG_SET_BITS_INT(invout, 0, 32, b);
16.       SC_STREAM_WRITE(inv_o, invout, tag);
17.       SC_STREAM_READ(word_i, wordin, tag);
18.    }
19.    SC_STREAM_CLOSE(word_i);
20.    SC_STREAM_CLOSE(inv_o);
21.    /// PROCESS_FUN_END
```

Figure 22 Process definition for an inverter

5.   SUMMARY

A model of computation (MOC) is a conceptual framework that allows a design to be specified, reasoned about, synthesized and tested. This chapter focused on models of computation that are used in digital systems – in particular embedded and reconfigurable systems. We discussed the various properties that characterize a model of computation. Then, we described the properties of a model of computation and attempt to present the terminology to allow a comparison between different models of computation. We continued with a discussion on classic models of computation. We detailed the similarities and differences from one another through their handling of different properties. Finally, we will give an overview of many of the popular MOCs used to design embedded and reconfigurable systems.

Chapter 5

A FRAMEWORK FOR SYSTEM SYNTHESIS

1.   INTRODUCTION

This chapter focuses on the development of a system compiler.   This compiler will accept a high level application description along with a target programmable device and create the necessary data to program the device to implement the application on a reconfigurable platform.   Figure 23 shows the design flow for the system compiler.

The application specification used for the system compiler is system design language based on the C syntax and semantics.   We choose C-based design language for several reasons.   First off, most programmers are familiar with the syntax and semantics of C.   Additionally, embedded applications tend to use a lot of legacy code, much of which is implemented in C.   Furthermore, it is easy to compile C to a wide variety of processors. This allows the framework to use existing highly optimized backend code generation for programming any microprocessing component that may be a part of the programmable device.   Additionally, it is easy to simulate designs written in C; you can simply compile the program and run it on the latest state of the art microprocessor.

Many system design languages use an extended from of C.   SpecC and SystemC [151-154] are two recent examples of system design languages that extend C with additional language constructs needed for embedded systems. C provides a good method for describing the behavior of functions; there are many methods for exploiting instruction level parallelism within C procedures.

However, C lacks the ability to provide the amount of task level parallelism needed in embedded systems. Additionally, it has no methods for specifying timing constraints that are required by most embedded applications. Some C concepts like recursive functions and pointers are extremely hard to implement in hardware. It may be possible to fully synthesis the full C language and some research efforts are looking into methods for doing this [155-157]. A restricted form of C can be transformed into a synthesizable behavioral HDL, which can be used as an entry point for logic level programmable devices like CPLD and FPGAs.

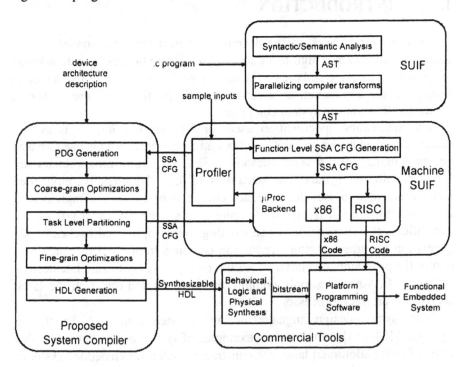

Figure 23 The design flow for the system compiler.

Languages for the specification of embedded systems is a rich and extremely active area of research [125, 128, 131, 150, 158-164]. Instead of creating yet another embedded system design language, the framework looks at design techniques that can be applied across a variety of systems design languages. Our framework focuses on the use of C as a design language because of its many benefits in embedded systems design. Of course, C is not an ideal language for every embedded application. We believe that the C language will continue its widespread use for embedded applications. Some possible future work based on this project could be the development of a system compiler that accepts multiple design languages. This would require an extensible intermediate representation (IR).

The system compiler discussed in this chapter is incorporated into the SUIF compiler infrastructure [165, 166]. The SUIF compiler has a front end, which provides *syntactic/semantic analysis* of several programming languages (C/Fortran/Java) into an *abstract syntax tree (AST)* intermediate representation (IR). Lam et al. have implemented many different high level *parallelizing compiler transformations* [167-170] using this IR. Using the SUIF infrastructure, Bondalapati et al. [171] have shown these methods are quite useful when compiling to reconfigurable systems; we can use these methods in a similar manner to create coarse level parallelism.

The Machine SUIF infrastructure supports back end optimizations targeting a variety of different microprocessors. Machine SUIF starts by transforming the SUIF AST into a *control flow graph (CFG)* with *static single assignment (SSA)* form. The infrastructure supplies the ability to produce C code from the SSA CFG form, which can be used for simulation purposes. Machine SUIF also provides a *microprocessor backend*. This can produces code for x86 and RISC microprocessors using the same extensible register allocation and instruction scheduling algorithms. Later in this chapter, we discuss methods of creating synthesizable HDL from the SSA CFG form.

Profiling is extremely important for the design of embedded systems. Profiling determines the important parts of the embedded application and allows us to focus our optimization efforts on these commonly occurring portions of the application. Machine SUIF provides a method for profiling applications. Profiling annotations can be placed anywhere on the SSA CFG IR. The x86 or RISC backend converts these annotations into function call hooks, which perform actions (e.g. writing to a file) when the code is run. This allows a wide variety of profiling.

The Machine SUIF infrastructure supports back end optimizations targeting a variety of different microprocessors. Machine SUIF starts by transforming the SUIF AST into a *control flow graph (CFG)* with *static single assignment (SSA)* form. The infrastructure supplies the ability to produce C code from the SSA CFG form, which can be used for simulation purposes. Machine SUIF also provides a *microprocessor backend*. This can produces code for x86 and RISC microprocessors using the same extensible register allocation and instruction scheduling algorithms.

The SSA CFG form is limited in many aspects. The basic block is often too small to exploit instruction level parallelism. Many different techniques have been proposed to extend the CFG into a form that allows for increased amounts of instruction level parallelism. Trace scheduling [14], superblock

[172], hyperblock [173] and PSSA [174] are some of the more successful techniques for enhancing the CFG form to generate more instruction level parallelism. All of these forms rely on a structured model of control flow, which often over specify the control flow constraints. It is difficult to detect coarse-grain parallelism using unless we move to a representation that shows only true control flow dependencies.

The *program dependence graph (PDG)* [175] efficiently expresses both control and data dependencies. The PDG is uses additional control dependence analysis to eliminate false control dependencies often found in the SSA CFG form. This allows more aggressive transformations and exploits coarse-grain parallelism. We give more detail on the advantages of the PDG in Section 4.

The *system partitioning* problem is a fundamental task for system design. At the system level, partitioning algorithms assign computing tasks to the system resource of the system architecture, while optimizing system performance metrics such as the execution time, hardware area (cost) and the power consumption. Some early works [176-179] investigate the hardware/software partitioning problem; it is difficult to name a clear winner [180]. The system partitioning problem is a generalization of the hardware/software partitioning problem. Partitioning issues for system architectures with reconfigurable logic components has also been studied [181-183]. These works assume a reconfigurable logic device coupled with a processor core in their partitioning problem.

Once we have decided the partitioning of the application across the components of the device, we must still generate all of the code needed to program the device. Once the system partitioning engine decides on the portions of the application that will be implemented on microprocessor components, it will transfer those parts of the application (in SSA CFG form) to Machine SUIF microprocessor backend to generate the code for these components.

The remaining portion of the application will undergo *fine-grain device dependent optimizations*. These optimizations include optimizing the communication between the various components. Finally, we perform *HDL generation*. The HDL generator creates all of the necessary information needed implement the given application on the programmable device. This *synthesizable hardware description language* is a mix of structural and behavioral HDL. The HDL compiler will utilize commercial tools for any behavioral, logic and physical synthesis techniques that may be needed for programming the device. For example, the synthesizable HDL can be fed into device vendor's commercial tools to create a *bitstream* for programmable logic device like an FPGA or CPLD. The synthesizable HDL and the microprocessor code (for those programmable devices with microprocessor components) are inputs to *platform programming software* provided by the manufacturer of the programmable device. This software creates a fully functional programmable device that implements the intended embedded application.

The rest of this chapter looks at the process of hardware compilation. First, we look at the hardware synthesis from a SSA CFG intermediate representation. This functional IR is restrictive for many parallelizing operations. We then describe the program dependence graph (PDG) as process level IR.

2.   HARDWARE COMPILATION

We start by introducing the basic ideas of our method of hardware compilation. *Hardware compilation* is the process of transforming an application level language to a form that can be synthesized onto a hardware component. A hardware component is any device that can be programmed by using an architecture-level behavioral description. The word "hardware" is an overloaded term; we define hardware as any device that can be completely programmed or fabricated from a hardware description language. There are many well-developed paths to take an architectural-level behavioral HDL description and use it to create a mask – the blueprints to fabricate an ASIC. Additionally, we would consider an FPGA hardware because it also has a set of tools to transform an HDL into a bitstream to program the component.

We call any optimization unit a *region*. A region is piece of the IR that we wish to focus our attention. We wish to transform a region into an *entity*. An entity is a set of input and output ports and a description of the relationship between the ports. An entity is a high-level entry point into the structural hardware synthesis flow. The *interface* of an entity is the input and output relationship of the entity to the outside world. A VHDL *architectural body* describes the relationship between the ports. It gives the functionality of the entity. The architectural body maybe defined behaviorally or structurally. Every entity is associated with at least one architectural body.

The input and output ports serve a variety of purposes. The main use of ports is for transferring data between entities, but we can also use ports to dictate the flow of control, a clock for the circuit, a reset, etc.

Assume that we wish to transform some region R to an entity E. The region R will be a portion of the IR and the entity E will be a synthesizable VHDL interface and corresponding architecture body.

Definition: A piece of data *d* is *generated* in region *R*, if and only if there exist an instruction *i* ∈ *R* such that *d* is the output operand of that instruction.

Definition: A piece of data *d* is *consumed* by a region R, if and only if there exist an instruction *i* ∈ *R* such that *d* is an input operand to that instruction.

The *data input ports* of the entity *E* would be any data that is consumed in the region *R* and generated by an instruction outside of the region *R*. A *data output port* is needed for any data that is generated by some instruction in the region and consumed by at least one instruction outside of the region.

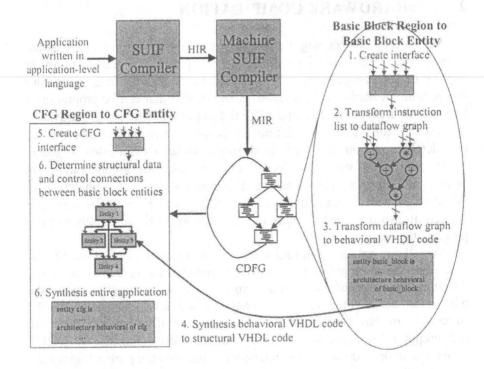

Figure 24 The flow from application specification to its hardware realization.

An entity can have many different architecture bodies. This allows different implementations of an entity. In addition, the architecture body can be modeled using a structural representation, behavioral representation or a mix of the two representations. We use a behavioral representation to model the entity describing a basic block region. The interconnections between the basic block regions are described using a structural representation. In effect, we are using a two level hierarchy for synthesis. The first (highest) level is a structural representation modeling the data and control interconnections between the basic block regions. The second (lowest) level is the modeling of the basic block regions as a basic block entity. The architecture body of a basic block entity is described using a behavioral representation. Each basic block entity is then synthesized using Synopsys Behavioral Compiler to yield a structural representation of the basic block. After the every basic block entity is synthesized, the entire design is realized as a two level hierarchical structural representation. We can then feed the design to a high level synthesis engine – we use Synopsys Design Compiler – to get to a logic level structural (gate level) representation. Then, we can hand off the design to any physical design tool to realize the final implementation of the application. Figure 24 shows a high level view of the entire flow, from application specification through system, architectural, logic and physical synthesis. The subsequent sections give in depth details on the entire flow.

3. BASIC BLOCK SYNTHESIS

We start the discussion of synthesizing the Machine SUIF IR [184] by describing the translation of a basic block to an entity. Equivalently, we describe the translation of a basic block region to a basic block entity. The CFG Node is the Machine SUIF description of basic block. The operations of the basic block are held in a linked list. The linked list enforces an ordering on the instructions. The first instruction in the linked list is executed before the second instruction, which is executed before the third instruction, and so on. There are many possible correct orderings of the instructions. In addition, many instructions can be executed in parallel. We will discuss the dependencies between the instructions in a later section. The instruction itself has an opcode that describes its functionality, a set of input operands and an output operand. We use the architecture independent opcodes from the SUIF Virtual Machine (VM) [185].

The operands are in the form of a variable symbol, a virtual register or an immediate operand. A *virtual register (VR)* (sometimes called a pseudo-register) is the notion of assigning a unique register to every point where data is generated. When assigning virtual registers, we have no regard for the "hard" registers of the final target microarchitecture. We assume that we have an infinite number of registers. The exact number and use of registers is decided during architectural synthesis. Similarly, a back-end for a processor microarchitecture (e.g. RISC) assigns the VRs to the hard registers in the register banks during the register allocation stage. By using VRs, the operands of the instructions show only the true dependencies – a read after write (RAW) dependencies. A *variable symbol* is a source generated variable or a compiler generated temporary variable [186]. An *immediate operand* represents a signed or unsigned integer immediate value.

Each instruction can be categorized as an arithmetic instruction, a control transfer instruction (CTI) or a memory instruction. An *arithmetic instruction* takes input operands and produces output operand(s). The opcode of the instruction dictates the functionality of that instruction. Some common arithmetic instructions are add, multiply, divide, shift, etc. The Machine-SUIF SUIFvm Library [185] describes all of the opcodes for the machine independent virtual machine used in our framework. All of the arithmetic opcodes from the SUIF VM have corresponding functionality in IEEE 1076.3 VHDL standard.

A control instruction determines the flow of the program. A jump instruction and a conditional branch are two examples of a CTI. In a microprocessor architecture, a CTI is any instruction that modifies the program counter. There is no analogue to a program counter when we synthesize to hardware (unless we choose to synthesis a controller that uses a program counter). The flow of control is dictated by a set of controllers. We will describe different methods for synthesizing control in a later section.

A *memory instruction* is any instruction that moves data between local memory and main memory. Local memory could be a register, register bank or scratch pad memory. The access latency for main memory can greatly affect the performance of the system. There are techniques (e.g. caching, prefetching) to reduce the effective latency of the main memory accesses. We assume that the read/write access times for main memory are on the order of a few cycles. A nice future study would be the effects of different memory access latencies on the design. Additionally, we assume that the main memory is a coherent, memory architecture i.e. every piece of data that we access from memory is the correct data and any data that we write to memory will be properly updated throughout the system.

The translation to VHDL is done from the CFG representation. As described in the previous section, each CFG Node of the CFG contains a list of instructions. First, we will describe how we translate such a list. The instructions we encounter will be of the type arithmetic instruction or memory instruction. There is at most one CTI instruction per CFG Node. We will describe how to handle the CTI instruction separately.

The entity for an instruction list and correspondingly, a CFG Node, has a set of input ports and output ports. The input ports are any VR operands that are consumed within the instruction list but not generated by an instruction in the instruction list. Additionally, if there is variable symbol that is consumed in the instruction list, there is an input port for that variable. There is an output port for a variable symbol if it is generated or, similarly, there is an instruction that has a destination operand that is a variable symbol. If a variable symbol is consumed and generated, there is an input/output port. Each of the variable symbol ports is directly connected to a single register that always holds the data for a variable symbol across the CFG. Additionally, there is an output port for every VR that is generated within the instruction list and consumed outside of the list.

If there exists a memory instruction, then we must add input and output ports corresponding to the interface of the main memory. The main memory can be any type of memory, (e.g. SRAM, DRAM, etc.) and located anywhere within the platform. The memory may be realized as an embedded RAM in an FPGA, hard core (ASIC) on a SOC, chip on a PCB platform or as any fully specified memory component of a platform library. Presently, we assume that the main memory interface is that of a scratch pad memory of the Synopsys DesignWare library. We use the synchronous write-port, asynchronous read-port RAM (DW_ram_r_w_s_dff) from the DesignWare library. Our framework allows for different memory interfaces with only minimal changes. The actual amount of main memory depends on the amount of data storage needed. The data storage would be the memory necessary to program the device, the appropriate input data, and storage for the output data. It is easy to envision that another component of the platform (e.g. a microprocessor) will use program this reconfigurable component and control the overall flow of the application on the platform. The reconfigurable component that we are synthesizing could be realized to a functional unit of some VLIW processor. We believe that design space exploration for main memory is an extremely interesting research topic, however it is out of the scope of this work.

We have defined the entity interface for the instructions in the basic block. Our next task is the creation of a behavioral VHDL architecture corresponding to the entity. We define a VHDL variable for every VR that is local. A VR is *local* if it is generated by an instruction in the basic block and never consumed by an instruction outside of the basic block. We now have defined every operand as either an input, output and input/output ports of the entity or as a variable in the architectural description.

The arithmetic instructions are translated directly to the IEEE 1076.3 specification. The VM "add" instruction is translated to "+". Likewise, the various shift, rotate, multiple VM instructions are translated into there IEEE 1076.3 counterparts. Not all of the VM instructions can be implemented using Synopsys Behavioral Compiler. Presently, Behavioral Compiler can only synthesize a subset of the 1076.3 specification. For example, Behavioral Compiler cannot synthesis an arbitrary division. Therefore, we change every division to a multiply instruction. This does not yield correct functionality of the compiled application, but we imagine that Behavioral Compiler will be able to synthesis the entire 1076.3 in the near future.

The memory instructions are handled by specifying the main memory interface.   We define   two   macros,   *main_memory_write*   and *main_memory_read*. The different memory instructions use these macros during the translation to VHDL. For example, the STR (store) operand from the SUIF VM has two input operands. One input operand specifies the data that is written to main memory. The second input operand gives the memory address or the location in memory to store the data. This instruction would use the main_memory_write macro. Additionally, I could associate the latency of the memory with different types of shared memory schemes.

Now that we have described a method for synthesizing a basic block, we must give ways to control the flow of execution through the basic blocks and the best manner to pass data between basic blocks.

3.1   Controlling the Flow of Execution

Now that we have translated every instruction in the basic block to its equivalent VHDL construction, we must provide a methodology to handle the flow of control.

To handle the control flow, we add an additional input port to every entity. We call this input port the EXECUTE port. When the EXECUTE port is set, the entity runs. When it is not set, the entity is in an idle state. It is easy to imagine that the idle state would put the hardware in low power mode.   There are some interesting low power hardware design methodologies, but any further discussion is out of the scope of the work. We must add a mechanism to direct the control flow i.e. a controller. We focus on two types of control – distributed control and centralized control.

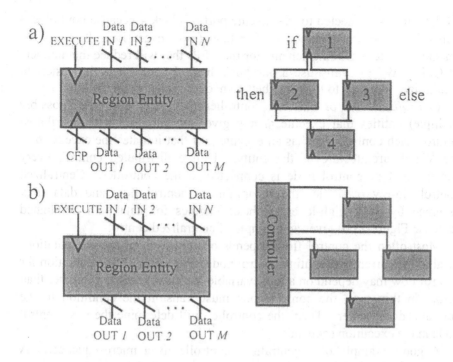

Figure 25 a) Distributed Control b) Centralized Control

Distributed control requires that the CFG Node entities control the path of execution. Each entity has a local controller that determines the next control node in the execution sequence. Therefore, there are direct control connections between control nodes. Additionally, each control node has a set of *control flow ports (CFP)*. There is a CFP for each of the different control nodes that may follow this node in execution. Equivalently, there is a CFP for each control edge of the CFG. A CFP connects to the execute port of other entities. Figure 25 a) illustrates a simple example of distributed control for an if-then-else clause.

Since the flow of control of a CFG uses local control transfer instructions (CTIs) to manipulate the control flow, it is natural to use a distributed control scheme. Each entity corresponding to a CFG Node determines the next block that should execute in the CFG. In essence, we must translate the CTI instruction of the CFG Node into a synthesizable VHDL description for the entity.

Consider Figure 25 a). The four basic blocks represent an if-then-else clause. Block 1, which corresponds to the "if" part, will transfer control to either block 2 or block 3 depending on the condition of the "if" clause. Therefore, block 1 has two CFPs – one to drive the "then" block (block 2) and one to drive the "else" block (block 3). Blocks 2 and 3 only need one

CFP, which is connected to the execute port of block 4 because both blocks will transfer control to block 4 after they finish execution. It is easy to imagine a better encoding scheme for the CFPs that will reduce the numbers of CFPs. We presently use a one hot scheme because it is the easiest to synthesize. We plan to look into better encoding schemes in the future.

Centralized control has one controller that determines the (possibly multiple) entities that execute at any given instant. As with distributed control, each control node has an execute port that initiates the execution of the VHDL architecture of the entity. Unlike distributed control, every execute port of control node is connected to the controller. Centralized control closely resembles the separation of control flow and data flow assumed by most high-level synthesis engines for data path dominated circuits. Figure 25 b) gives an example of centralized control

Most often the control flow depends on the result of the computations local to the currently executing control node. For example, the condition for control flow may depend on a local variable, e.g. that variable is greater than zero. In this case, the control node must transmit the condition to the centralized controller. Then, the controller will determine the next control node in the execution sequence.

A good example of a centralized controller is a microsequencer. A microsequencer translates an instruction into the control signals that must be asserted in order to properly execute the instruction. In our case, the instructions would be the basic blocks themselves as well as the needed control instructions. The CFG would then be "compiled" into these basic block instructions and the necessary control instructions to direct the CFG. The microsequencer would translate these instructions and assert the proper control signals (execute ports) to enable the execution of the CFG.

Centralized control has the advantage that one parameterized controller can be created and then used for any application. However, the centralized controller must be generalized to handle any type of control scheme that is given to it. Therefore, the centralized controller is often overdesigned for a given application.

As an aside, the way that we currently implement the control allows only one CFG Node to execute at a time. However, it is possible to extend this format to include a group of nodes that execute in parallel. We simply allow an entity describe a set of CFG Nodes, instead of a single CFG Node. The problem of selecting a suitable set of CFG Nodes to group as an entity is similar in nature to the hyperblock selection problem, which was first introduced for VLIW architectures [187]. Due to limited space, we will not discuss this subject further, only to say that it is possible to allow multiple CFG Nodes to execute in parallel and it is a good direction for future work.

To measure the effectiveness of the two control flow schemes, we examined a set of DSP functions. DSP functions typically exhibit a large amount of parallelism making them ideal for hardware. The DSP functions were taken from the MediaBench test suite [188] (see Table 2). The files were compiled into CDFGs using the SUIF compiler infrastructure [165] and the Machine-SUIF [166] backend. Each of the functions was given to our framework, which produced synthesizable VHDL code. Then, they were synthesized using the Synopsys Behavioral Compiler for architectural synthesis followed by the Synopsys Design Compiler for logic synthesis.

Table 2 MediaBench benchmark description

Application	C File	Description
mpeg2	getblk.c	DCT block decoding
adpcm	adpcm.c	ADPCM to/from 16-bit PCM
epic	convolve.c	2D general image convolution
jpeg	jctrans.c	Transcoding compression
rasta	fft.c	Fast Fourier Transform
rasta	noise_est.c	Noise estimation functions

Function	Name	Num Control Nodes
adpcm_coder	adpcm1	33
adpcm_decoder	adpcm2	26
internal_expland	convolve1	101
internal_filter	convolve2	101
compress_output	jctrans	33
decode_MPEG2_Intra_Block	getblk1	75
decode_MPEG2_Non_Intra_Block	getblk2	60
decode_motion_vector	motion	15
FAST	fft1	14
FR4TR	fft2	76
det	noise_est	12

We wish to compare the distributed control scheme with the centralized control scheme. We have fully implemented the aforementioned distributed control scheme into our framework. For a comparison, we will use a microsequencer controller [189] as a centralized controller. The microsequencer was synthesized with ability to handle 25 operations.

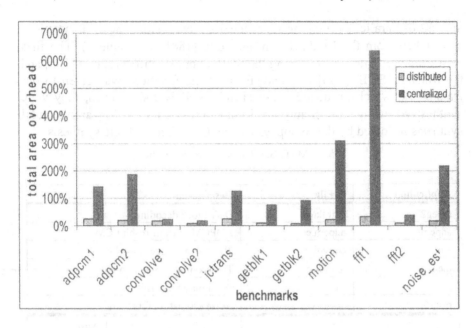

Figure 26 The area overhead comparison of the distributed and centralized schemes

The MediaBench functions were first synthesized without any global controller to get a baseline area result. Then, we added the distributed scheme and synthesized the functions again. All of the benchmarks were synthesized using the pruned algorithm for data communication. The percentage area overhead is shown in Figure 26. The average overhead for the distributed scheme is 18.6%. To get the approximate area results for the centralized controller, we synthesized the microsequencer [189] assuming that each CFG Node has one control signal and the average cycle time per node is 7 cycles. The microsequencer adds a constant fixed amount of area to each of the functions. This overhead is quite large; for the function FAST, the controller overhead is over 6 times that of the datapath of the function.

While the centralized controller adds a fixed amount to each function, the area overhead of the distributed scheme is linearly related to the number of nodes in the function. This explains the similar area overhead of the distributed and centralized scheme when considering the benchmark internal_expand (101 control nodes). This also explains the immense overhead of the centralized controller on the benchmark FAST. FAST has only 14 control nodes. Therefore, as the number of control nodes increases, the area overhead of the two schemes becomes similar.

The area of our synthesized centralized controller does not account for interconnect area as well as the connections from the controller to the control nodes. Furthermore, we did not account for the extra memory components needed to store the control flow instructions for the function. Despite this fact, the distributed controller has a smaller area overhead as compared with the centralized controller. Additionally, the latency of a function with a centralized control will be larger than that of the distributed controller. The centralized controller must decode the current control instruction after each control node executes. The distributed scheme can often calculate the next control node execution in parallel with the other instructions of the basic block. Therefore, we conclude that the distributed scheme is the best control scheme for our framework.

3.2   Data Communication Schemes

In addition to determining the type of control for the CDFG, we must determine the method of data communication between the regions. Once again, there is a centralized and distributed method of data communication. A *centralized data communication scheme* passes the data through a centralized storage area such as a register bank or RAM block, depending on the amount of data. This allows a memory hierarchy scheme where data can be cached and large amount of data can be accessed by the CDFG.

A *distributed data communication* scheme passes the output data from the currently executing entity directly to the inputs of the control node(s) that might need the data later. The output of the entity may connect to multiple other entities including itself (in the case of loops).

As with most of engineering decisions, there are benefits and drawbacks to consider for each of the communication schemes. The distributed data communication scheme is simple to implement, as you do not have to worry about interfacing to bus and memory protocols. Additionally, the distributed scheme has direct connections, meaning that the communication between control nodes will occur quicker compared to the centralized scheme; data passing does not involve writing to and reading from a central memory bank. Yet, the centralized scheme allows a sharing of resources. The distributed scheme will have more connections (interconnect), many of which will not be active at a particular time leading to a waste of communication resources. Furthermore, the increased connectivity between control nodes may have a negative impact on the circuit's area.

We study different methods to minimize data communication in Chapter 8. In our framework, we have implemented a locally distributed, globally centralized data communication architecture. The values that are generated and used by the CFG are stored in local registers. This allows quick read and write access for the values. Any value that is generated outside of the CFG and used by the CFG is stored in a global memory. In such a case, the

CFG will make a memory call to a global memory component. In addition, any value that is generated in the CFG and used outside of the CFG will be written to global memory. Global memory could consist of many main memory components. Then, we would need to determine the best way to distribute the data across the memory components. Multiple memory components may enhance memory performance, though this would require additional optimizations to determine the best manner to distribute the data. We make the simplifying assumption that there is only one global memory component. Hence, we use the terms global memory and main memory interchangeably.

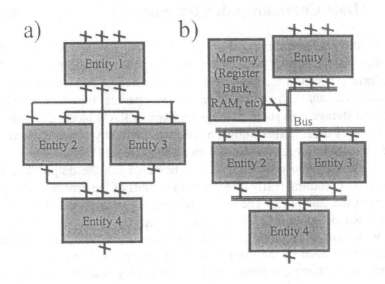

Figure 27 a) Distributed data communication b) Centralized data communication

Our scheme requires that we determine the *local values* – values that are both generated and used locally to the CFG and the *global values* – any value that is generated outside of the CFG and used within the CFG or generated within the CFG and used outside of it. We can leverage Machine-SUIF to determine the local and global values.

A global value is any memory access instruction as translated by Machine-SUIF into the Machine-SUIF SUIFvm architecture. This allows the use of pointers, though we assume that the global memory will handle any arbitrary address that we may give it. A formal analysis of the possible pointer values in order to restrict the address space would be quite beneficial, as it would reduce the size of the memory giving greater efficiency in terms of silicon usage as well as reducing the memory latency.

We refer an interested reader to the many recent works of Rinard et al. [190-198] for a techniques and other references for pointer and alias analysis. Machine-SUIF will also translate any array accesses as global values. It is possible to treat the entire array as a set of local values, though our framework treats all arrays as global values.

The local values are translated to virtual registers (VR) by Machine-SUIF. In addition, we treat parameters and static variables as local values. We create a register for each variable and assume that the system will load the initial values into those registers before executing the hardware that we synthesize. Since we are synthesizing each basic block separately, we must determine what local values that the basic block generates and uses. In addition, we must create a way to transfer the local values between basic blocks. A simple method for handling the transfers would create a register for each local value. Then, a basic block that uses the value would be connected to the register that holds the values and retrieve the value from the corresponding register. Also, any basic block that generates a value would be connected to a register. With the reuse of values and different paths of control, it is possible that a register is connected to a basic block that will never use the value. This would require increase the number of the number of multiplexors and control logic needed to handle accesses to a register. In turn, this increases the area and latency of the registers. We explain a scheme to handle this problem in Chapter 8. We defer the exact determination of passing data through the CFG to that chapter.

4. PROGRAM DEPENDENCE GRAPH

The *program dependence graph (PDG)* [175] efficiently expresses both control and data dependencies. The PDG is uses additional control dependence analysis to eliminate false control dependencies often found in the SSA CFG form. This allows more aggressive transformations and exploits coarse-grain parallelism.

The PDG is an advanced compiler IR that has shown many benefits. It has been shown to be able to efficiently handle interprocedural analysis [199] and can be transformed to maintain the static single assignment property [200-202]. The PDG eliminates the artificial order imposed by the AST and CFG. It allows for both coarse-grain and fine-grain parallelism optimizations.

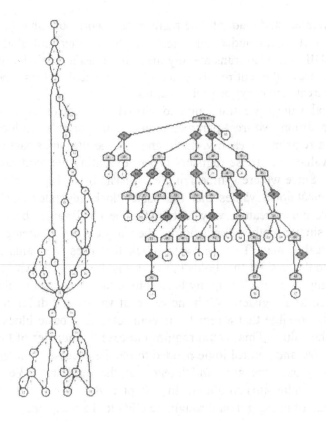

Figure 28 The add_block function from the MPEG2 decoder application in the MediaBench
test suite. The left graph is the CFG representation for the function while the right graph
is the PDG representation of the function. The circles denote the basic blocks of the
function. The diamonds are control statements and the remaining symbol ("house"?)
denotes a PDG region node.

The PDG supports high level transforms available in the AST. For
example, scalar expansion and array renaming are both possible in the PDG
form. The PDG also enables node-splitting, which duplicates the PDG
nodes and divides its edges between two copies to break dependence cycles.
Moreover, the AST cannot directly perform loop interchange and
vectorization; the AST must perform additional if-conversion analysis. The
PDG can perform both of these optimizations because the PDG edges
express control and data dependencies uniformly. Additionally, the PDG
supports traditional program transformations such as common sub-
expression elimination and constant expression folding [175]. The PDG is a
superior form to CFG based forms because these forms use a structured
model of control flow. The PDG shows only true control and data

dependencies. Figure 28 illustrates the difference between the CFG and PDG forms for a function from an MPEG2 decoder from the MediaBench test suite [203]. You can see that the PDG form produces fewer control dependencies and allows for more concurrent operation of the application.

The PDG is widely used in parallel computation and is starting to be seen as an IR for embedded system languages. Edwards [204] uses the PDG as an IR when compiling Esterel programs into hardware. Esterel programs are control dominated and circuit generation from the PDG is compact and better than circuits generated from a CFG form. Ramasubramanian et al. [205] use the PDG form to analyze loops in the synthesis of reconfigurable systems. We plan to study sensor networking applications, which tend to have a complicated mix of control and dataflow.

Creating the PDG form requires a large amount of dependence analysis. This increases the compilation time of compilers that use the PDG representation. Most current microprocessor compilers do not use the PDG form. The functional level transformations (done on the CFG form) exploit enough parallelism to keep the processors resources busy. Today's microprocessors are optimized to exploit instruction level parallelism. However, future programmable devices, which contain multiple processing components, need techniques for exposing coarse-grain task level parallelism. The PDG has much opportunity to exploit support coarse-grain parallelism for these future devices.

5.   RELATED WORK

There are many projects looking at design environments for reconfigurable devices. We will highlight some of these projects and describe their relevance/differences with our work.

The RaPiD compiler [105] forces the programmers to specify coarse-grain parallelism using a special architecture specific language called RaPiD-C. The PDG allows for automated methods for exploiting more task level parallelism.

The DEFACTO project [171] uses the abstract syntax tree (AST) intermediate representation to perform high level transformations like scalar and array optimizations. The AST is a high level representation, so it lacks a connection with the targeted programmable architecture. Nevertheless, the techniques they use are effective at exploiting parallelism and we plan to incorporate their techniques into our framework (they implemented their algorithms using SUIF infrastructure).

The RAW [109], Garp [29], Sea Cucumber [206], and CASH [207] compilers focus on a CFG-based intermediate representation. RAW uses

SSA CFG form. Garp uses hyperblocks to further exploit instruction level parallelism. Sea Cucumber and CASH use a more advanced form similar to PSSA. The PDG form that we use is allows for coarse-grain task level parallelism as well as instruction level parallelism as provided by these forms. These methods can be used for fine-grain optimizations on the PDG in conjunction with of the proposed coarse-grained methods.

Edward's recent work [208] uses the PDG, but it focuses on applications written in Esterel. Esterel is used to describe extremely control dominated applications. Our work focuses on applications that have a complicated mix of control and data flow.

The Cameron project uses a design specific language called SA-C [209]; they translate SA-C into an intermediate form called the data dependence and control flow graph (DDCF). The DDCF is a similar form as the PDG. Currently, they perform a one-to-one mapping of application resources to an FPGA. They do not perform automated partitioning across different programmable platforms. Furthermore, they require the designer to use their SA-C language, which is primarily targeted towards image processing applications.

There is a good amount of work focusing on the creation of a hardware compiler e.g. [29, 151, 154, 206, 209-214]. Some of these works create new languages for synthesis to hardware [151, 154, 209, 210]. We feel that it is better synthesize from an IR – rather than adopting a specific language, which allows the possibility of using our framework to compile many different system design languages.

Many works [29, 206, 211-213] perform synthesis from an IR. Our work takes a unique approach to hardware synthesis by allowing resource sharing within an entity. To the best of our knowledge, the previous works perform a direct one-to-one mapping of operations to resources. This can potentially cause a huge waste of area, which is unacceptable when compiling to ASICs. Even when we are compiling to FPGAs, it is advisable to limit the area usage so that we can potentially use a smaller FPGA in the system. Furthermore, one could perform some sort of configuration caching where multiple applications share the FPGA; therefore we would want to limit the area of the applications to reduce the large reconfiguration overhead.

6.   SUMMARY

It is evident that we must move to a higher level of abstraction in order to handle the complexity of designing new computing systems. We have presented a framework that can transform a medium-level internal representation (MIR) of a compiler into synthesizable VHDL. Our framework integrates the SUIF compiler with Synopsys Behavioral Compiler. This realizes a complete path from a system design language to a hardware description.

We studied two important steps in system synthesis – data communication synthesis and control synthesis. We described the locally distributed, globally centralized scheme of data communication in our framework. We examined the use of different SSA algorithms to minimize the local data communication at the system level (within the compiler) and showed that final synthesized area results are most favorable using the pruned algorithm. Furthermore, we explored two different types of control schemes – distributed and centralized. We showed that the centralized scheme incurs a high area cost due to its general nature. We demonstrated that our new distributed control architecture is more area efficient and argued that it gives a better latency. Finally, we described a new representation called the program dependence graph and described the advantage of this representation over other IRs.

In this chapter, we attempted to give the reader an understanding of the underlying framework that we assume throughout the rest of the work. The framework is not complete. We have tried to convey the shortcomings and the assumptions that we make. We have attempted to make a standard framework that can be extended as well as enhanced using the SUIF infrastructure. The framework conforms to the SUIF 2 methodology of creating independent passes so that they can be replaced or interleaved to suit the user. The framework is still in the research stage; it is continually being improved by a group of researchers at UCLA and UCSB.

Chapter 6

HARDWARE/SOFTWARE SYSTEM PARTITIONING

A popular means of accelerating an application's execution is to extract portions of it for realization in hardware. In the most ideal performance situation, the entire application would be implemented as hardware logic rather than as a software product, as hardware execution is many times faster than software. Unfortunately, the cost and size constraints of practical hardware systems make it impossible to realize large applications as hardware alone. Therefore, designers frequently consider solutions comprised of both software and custom hardware, working together to meet a set of performance constraints. The exploration of the system design space formed by combinations of hardware and software components is referred to as *hardware/software codesign*.

The foundations of codesign lie in embedded system design research, in which designers have often augmented a microprocessor with custom hardware to achieve boosts in system performance. The proven performance potential of these hardware/software systems, combined with the rapid evolution of powerful reconfigurable logic platforms, has sparked recent interest in architectures which combine a general-purpose processor with a reconfigurable coprocessor. Several industrial [215-217] and academic [25, 26, 109] platforms have been built combining microprocessor and reconfigurable logic on a single chip.

The ability to quickly and easily program hybrid hardware/software systems is of key importance if they are to be widely adopted. Ideally, hybrid application design should approximate the ease of application design for conventional microprocessor platforms. Without simple design methodologies, system performance would have to be considered against an unfavorably long time-to-market, as programmers would be bogged down in overly complicated implementation and verification cycles. Therefore, platform designers typically seek to enable specification of hardware/software systems from a single description, usually written in a high-level programming language [218-222].

The division of a system specification into hardware and software components is referred to as *hardware/software partitioning*. A partitioning of a system description impacts many aspects of the solution's quality, including but not limited to system execution time, total hardware area and cost, and power consumption. Unfortunately, as system partitioning typically occurs at a very high-level in the design flow, system characteristics resulting from a particular partitioning decision can only be estimated. The estimated characteristics of a partitioning are used to evaluate a cost function, which rates the overall quality of a given partitioning. Thus, the design space is explored in the quest for the highest quality system partitioning (i.e. the hardware and software partition that minimizes a cost function for an arbitrary set of constraints).

The rest of this section will provide an overview of hardware/software partitioning, and its connection to embedded and reconfigurable system research. We start by formally expressing the classic partitioning problem and then formulate Hardware/Software Partitioning problem mathematically, and then discuss the solutions and problem refinements that have been composed over the last decade.

1.   THE GENERAL PARTITIONING PROBLEM

Partitioning is a fundamental CAD optimization problem. It is used at almost every level of abstraction during the synthesis of a digital system. The partitioning problem attempts to take a set of connected modules and group them to satisfy a set of constraints and optimize a set of design variables. During physical synthesis, partitioning is used during the floorplanning and placement tasks. In this case, the modules are gates that are connected by nets. We wish to create a partitioning such that highly connected gates are in the same partition. This reduces the amount of interconnect delay caused by the wires between the gates. As we move to higher levels of abstraction, the modules become larger; they move from standard cells to macro cells (logic level) to blocks (architecture level). Partitioning at higher levels of abstraction will impact the system performance in a more drastic way. A bad partitioning at the architecture level is hard to correct at lower levels of abstraction. Furthermore, the interconnect delay at the higher levels of abstraction is more pronounced. The delay between two blocks is orders of magnitude larger than the delay between two standard cells. Hardware/software partitioning lies at one of the highest level of abstraction for the design of digital system. A good system partitioning is essential for the overall quality of the circuit.

A *k-way partitioning problem* is the basis for any variant of the partitioning problem. Given a set of modules $M = \{m_1, m_2, \ldots m_n\}$, the k-way partitioning problem finds a set of clusters $P = \{p_1, p_2, \ldots p_k\}$ such that

$$p_i \subseteq M \text{ for } i \rightarrow 1 \text{ to } k, \bigcup_{i=1}^{k} p_i = M \quad p_i \cap p_j = \varnothing \text{ for } i \rightarrow 1 \text{ to } k, j \rightarrow 1 \text{ to } k \text{ and i}$$

$\neq$ j (i.e. the clusters are mutually disjoint). The partitioning solution must satisfy a set of constraints and optimize an objective function. The objective function and constraints are subject to the type of problem we are trying to optimize. When $k = 2$, the problem is called *bipartitioning*.

The partitioning must be evaluated using some sort of estimation function. The estimation function $E(P) = DP(dp_1, dp_2, \ldots, dp_q)$ takes a partition P and returns a set of design parameters DP. The design parameters are properties of the circuit such as the area, power, throughput, latency, etc. The estimation function is used to tell us if the constraints are met and also to evaluate the optimization function. The estimation function is extremely important, especially at the system level. The ideal estimation function would run through the entire synthesis flow. Unfortunately, this is an extremely time consuming task. A large design can take days to fully synthesize. Obviously, this is unfeasible, especially if we wish to evaluate a large number of partitions. Therefore, the quality of the partitioning directly depends on the precision of the evaluation function.

There are many methods and heuristics to solve the partitioning problem. Alpert and Kahng [223] give an excellent survey on formulations and heuristics to solve the netlist partitioning problem. The next section gives a detailed formulation of the Hardware/Software Partitioning problem.

2. FORMULATION OF THE HARDWARE/SOFTWARE PARTITIONING PROBLEM

One of the clearest mathematical formulations of the Hardware/Software Partitioning problem was presented in [224]. The following is a slight modification of the problem as presented in their work.

Given: a set of tasks $T = \{t_1, t_2, \ldots t_n\}$ that comprise the complete functionality of the system, as well as a set of performance constraints $C = \{C_1, C_2, \ldots C_m\}$, where $C_i = \{G_i, V_i\}$, $G_i \subset T$, and $V_i \in Real > 0$. (V_i is a time constraint on the maximum execution time allowed for all tasks in task group G_i. Each task t_i can represent system functionality at nearly any granularity: as coarse-grained as an entire program or as fine-grained as a single micro-operation on a processor.)

Define: a hardware/software partition as $P = \{H, S\}$, where $H \subset T$ and $S \subset T$, and $H \cup S = T$, and $H \cap S = \varnothing$. The hardware size of H is defined as the area needed to implement the tasks in set H in hardware. The performance of the system is defined as the total execution time of a set of tasks G is defined as the total execution time of the tasks in G, or

$$Performance(G) = \sum_{t_i \in G} execution\_time(t_i).$$

A *performance-satisfying partition* is a P such that $Performance(G_i) \leq V_i$ for all $i = 1...m$.

Using the inputs and definitions provided, the original Hardware/Software Partitioning problem is to find a performance-satisfying partition $P = \{H, S\}$ such that the hardware size of H is minimal. With such a partition, the system would be able to satisfy its performance constraints (for which additional hardware beyond the microprocessor was required) with a minimal amount of coprocessor area. The problem is a subset of the General Partitioning problem for a system containing one or more additional ASIC or FPGA resources (as the set H could be subdivided into functionality across many pieces of hardware).

Over time, the goals of the Hardware/Software Partitioning problem have changed depending on the design situation. For instance, a general computing system (for which there are few or no performance constraints) might contain a RISC microprocessor and a fixed-size FPGA coprocessor. On such a system, the nature of the problem would be different, as the design constraint would be hardware size, and the goal would be to minimize execution time of T. We define an *area-satisfying partition* to be one in which the hardware size of $H \leq A$, where A is a given maximum hardware size. The Modified Hardware/Software Partitioning problem would be to find an area-satisfying partition $P = \{H, S\}$ such that *Performance(T)* is minimal. Other systems seek to minimize other characteristics, such as power consumption of the system [225]. These modifications are usually represented as straightforward modifications of the mathematical formulation presented here.

The Hardware/Software Partitioning problem is NP-complete, as it requires the exploration of a design space that is exponential in size (relative to the number of tasks in T). (The Hardware/Software Partitioning problem can be restricted to the Precedence Constrained Scheduling problem (PCS), by setting the amount of time of execution of hardware and software tasks to be the same. PCS finds an m-processor schedule for an ordered set of tasks subject to precedence constraints such that deadline D is met. PCS is NP-complete, and therefore so is Hardware/Software Partitioning [11, 226].) The time of determining a partitioning solution is therefore proportional to $|F|$, and therefore the granularity at which tasks are considered will largely impact the performance of a partitioning heuristic. However, as task granularity becomes coarser and the design space is decreased, design quality is potentially sacrificed. In an extremely coarse-grained case, where only two very large tasks are considered ($|F| = 2$), the partitioning of the system would be very easy. However, considering the partitioning of subdivisions of each task might have led to higher quality system.

The hardware/software partitioning must be performed early in the design cycle, as the functionality residing on a piece of hardware must be determined before the hardware can be synthesized. As a result, the design characteristics (including hardware size and performance of a partition P) must be estimated at the earliest stages of design. The accuracy of a given partitioning is as only as good as the estimation of the design parameters for that partitioning. The estimations provide insight into the quality of a solution. They must be fast, to keep the execution time of the partitioning heuristic within reasonable bounds.

Therefore, the quality and performance of the partitioning solution depend largely on the granularity at which tasks are considered, the performance and accuracy of design estimators, and the performance and effectiveness of the search heuristic utilized. These factors make the Hardware/Software Partitioning problem very complex, and difficult to solve for a general system.

3. INITIAL MODELS AND SOLUTIONS FOR HARDWARE/SOFTWARE PARTITIONING

Early investigations into codesign for embedded systems frequently mentioned characteristics that a good system partitioning would possess, although the researchers made no attempt to automate the solution to the partitioning problem. Researchers at Carnegie Mellon University [227] described a type of system codesign that extracted portions of application software for hardware realization, directed at acceleration of the application. They developed a model for system level simulation and synthesis, including a transformation capability allowing the generation of design options. Their proposed partitioning would take place at a coarse-grained task level rather than at the operation level, although no specific algorithm was describe in that work. They characterized the quality of a function level partitioning by three criteria: the impact of the partition on total system execution time, the difference in execution time between a given task in hardware as opposed to software, and the total cost of the custom hardware required to realize a task. The Ptolemy group at UC Berkeley also developed an early codesign methodology and framework [228], but in this work the designer was required to partition the design manually. (Subsequently, Kalavade and Lee have developed techniques to automate design partitioning, as well as to solve an extended form of the problem [229].)

Gupta and DeMicheli developed one of the earliest algorithms to automate a search of the design space for a hardware/software partitioning [230]. Their algorithm operated on an operation-level granularity, and was greedy in nature. Starting with an initial solution where all functionality was implemented in hardware ($H = T$, $S = \varnothing$), their partitioning heuristic selected operations for movement into software based on the cost of communication overhead. Movements were not taken if they did not improve the cost of the current system partition, or if they violated the given set of imposed timing constraints. The algorithm iteratively improved the partition until no cost-improving move could be found. The decision to start with an all-hardware partition was chosen to ensure that no infeasible design was considered. (An initial all-hardware solution is assumed to satisfy the performance constraints of the system. From this point, only moves which continue to satisfy performance constraints are accepted.) An assumption of the Gupta/DeMicheli algorithm is that most inter-function communication will happen between successive operations. Therefore if an operation is moved from hardware to software, its successor operations are given priority consideration as possible movement candidates.

Due to the greedy nature of the Gupta/DeMicheli algorithm, and its resulting inability to make short-term increases in system cost to achieve long-term cost minimization, it was easily trapped in a local minimum. Therefore, sub-optimal partitioning solutions were typically arrived at. Additionally, the algorithm frequently created very costly hardware consuming many resources, due to the initial partition being an all-hardware solution. (If cost-reducing movements from hardware to software were exhausted, the remaining design would be accepted. Typically the algorithm completed before the hardware area was reduced to a pragmatic quantity.) To overcome the limitations of this initial development, hill-climbing heuristics such as simulated annealing were subsequently explored.

Ernst and Henkel developed a hill-climbing partitioning heuristic that sought to minimize the amount of hardware used, while meeting a set of performance constraints [176]. Their work operated on the basic block level of functional granularity, slightly coarser than the work of Gupta and DeMicheli. Like the Gupta/DeMicheli algorithm, they started with an initial partitioning that was improved on subsequent iterations. However, to escape convergence to a local minimum, they utilized simulated annealing to explore design cost. Unlike greedy heuristics, simulated annealing often accepts changes which decrease the quality of a design, in hopes of achieving a more optimal final design. The process of simulated annealing is controlled by a temperature parameter, which begins at a high value and decreases as the system "cools" and stabilizes. Initially, during high temperature, moves which increase system quality are always accepted, and moves which decrease design quality are accepted randomly. As the temperature approaches zero, only moves which decrease system costs are accepted. Ernst and Henkel began the process with an all-software partition, seeking to minimize hardware costs by starting with less hardware. Obviously, their initial partition generally violated the performance constraints of the system. In order to prevent annealing before a performance-satisfying partition has been reached, they used a heavily weighted cost function that provided high penalties for violating runtime constraints. This choice proved effective in minimizing hardware costs. To provide performance enhancement estimates for hardware implementation, Ernst and Henkel utilized simulation and profiling information to determine the most frequently executed and computationally intensive regions of functionality.

Like Ernst and Henkel, the early work of Peng and Kuchcinski [231] also utilized simulated annealing in order to approach a more optimal partitioning. Their algorithm provided a more general multi-way partitioning (as opposed to the two-way partitions of other works), which enabled a decomposition of system functionality into a number of clusters (each resulting in a different hardware or software component). They modeled the system functionality as a Petri Net, which was then used to construct a graph representing the components and their communication. The weight of each node represents the cost of building this component, and the weight of an edge between two nodes represents the cost of implementing the data connection. The goal of their partitioning was to decompose the graphs into a set of sub-graphs such that the sum of the weights of all cut edges is minimized, while the total weights of the sub-graphs are balanced. Their work, although promising in its application to general system partitioning, lacked any direct representation of system performance, and also described no specific hardware minimization technique.

Vahid, Gong, and Gajski [224] built upon the work of their peers by developing a technique that partitioned a set of very coarse granularity tasks into software and a minimal amount of hardware. The task granularity was on the order of statement-level control constructs, such as loops and function bodies. Large variables were also considered as "tasks" to be partitioned. In this work, the goal of the partitioning is closely aligned with the formulation of the Hardware/Software Partitioning problem mentioned earlier. They describe a partitioning algorithm named *PartAlg*, based on a hill-climbing heuristic (much like the work of [176, 231]), which takes an input set of performance constraints and a maximum hardware size C_{size}. Every design possibility is weighed with a cost function equal to the weighted sum of runtime performance and hardware size violations. In other words, every violation of a performance constraint, or any violation of input maximum hardware size C_{size} contributes to the cost of a design. The assumption is that there exists a threshold input hardware size C_{size} at and above which the design cost of *PartAlg*'s solution will be equal to zero (as performance constraints will be met, and the hardware area constraint will be large enough that the design will satisfy it). Under this assumption, their solution is a nested loop algorithm. The outer loop is a binary search of the solution cost of the inner loop's *PartAlg* for various values of input area constraint C_{size}. The idea is to find the smallest C_{size} at which the partitioning returns a zero-cost design; this C_{size} will be the minimal hardware partition satisfying the performance requirements of the system. Although their heuristic reduced total hardware area compared to the Gupta/DiMicheli and hill-climbing approaches, the execution time of the algorithm suffers due to the execution of an iterative improvement algorithm roughly $log(n)$ times. The coarse granularity of the operations considered helps reduce the execution time, but may also lead to less optimal design results.

4.   REFINED AND ENHANCED HEURISTICS

Subsequent rigorous heuristics to model and solve the Hardware/Software Partitioning problem are typically refinements of the previously mentioned approaches. In this section, we briefly outline some of the more recent improvements to the techniques described above.

The translation of the Hardware/Software Partitioning problem into a set of integer programming (IP) constraints was described in the work of [232]. The extraction and creation of coprocessing hardware is broken into two stages. The first phase solved the traditional Hardware/Software Partitioning problem by estimating the schedule time for each functional node, and the second phase generated a correct scheduling for hardware/software-mapped nodes. The IP model devised by the authors is one of the most flexible partitioning heuristics presented, as it accounts for multi-coprocessor technologies as well as hardware sharing and interface costs. Additionally, a high-level synthesis schedule is produced for the functional nodes, guaranteeing system execution-time constraints. This is one of the most optimal and complete solutions proposed, and the algorithmic execution remains reasonable due to the utilization of better metrics (resulting in fewer iterations).

Kalavade and Lee, whose previously cited work on the Ptolemy project contained no automated partitioning heuristic, introduced the Global Criticality/Local Phase (GCLP) algorithm to solve the two-way partitioning problem [229] for tasks of moderate to large granularity. The authors note that two possible objective functions could be used to map a task to hardware or software: minimization of the execution time of that node, and minimization of the solution size (hardware or software area) of the node's implementation. To this end, the authors devise a global criticality measurement, which is re-evaluated at each step of the algorithm to determine whether time or area is more critical in the design. As the list of functional tasks is traversed, the global criticality measurement is checked to determine the current design requirement. If time is critical, the mapping minimizes the finish time of the task; otherwise the resource consumption of the task is minimized. In addition to the global system requirements, local optimality is sought by classifying each task as either an extremity (meaning it consumes an extreme amount of resources), a repeller (meaning the task is intrinsically preferred to either have a software or hardware implementation), or a normal task. This classification of each task, and its weighty consideration in the choice of hardware or software mapping, represents the local phase of a given task. The running time of the GCLP algorithm is extremely efficient $(O|N|^2)$, and the partitions it determines are no more than 30% larger than the optimal solution (as determined by an ILP formulation). (The work of Kalavade and Lee also formulates and heuristically solves an extension of the Hardware/Software Partitioning problem, which simultaneously determines whether a task should be implemented in hardware or software, as well as the best speed/area hardware implementation tradeoff for the task. Since the focus of this work is largely an extension of GCLP to hardware synthesis of a high-level task, it is beyond the scope of this section.)

Just as Kalavade and Lee incorporated a dynamic performance metric into their partitioning decision, Henkel and Ernst extended their previous work on the COSYMA environment [176] to incorporate a dynamic functional granularity [233]. Whereas their earlier work had been limited to basic block granularity, the authors' new partitioning method allowed the dynamic clustering of fine-grain tasks (at the basic block or instruction level) into larger units of operation (as large as the procedure/subroutine level). The rationalization for having a flexible functional granularity are that large partitioning objects should contain whole control constructs (in the form of loop bodies or procedures), and that only a few moves should be necessary (between hardware and software) to determine a good partition. The authors' innovation amounts to a hierarchical search of the design space, and the fast retrieval of a good solution by applying this modification to their earlier simulated annealing approach.

As an alternative to the simulated annealing-based iterative-improvement heuristics explored by their contemporaries, Vahid and Le extended the Kernighan-Lin (KL) circuit partitioning heuristic to explore the design space of Hardware/Software functional partitioning [179]. The chief advantage of the KL heuristic is its ability to overcome local minima without making excessive numbers of moves. The basic strategy of KL is to make the least costly swap of two nodes in different partitions, and then to lock those nodes. This continues until all nodes are locked. The best partition *bestp* is selected from this set. All nodes are subsequently unlocked, and the previous *bestp* becomes the starting point for the next set of node swaps. This swapping, locking, selection of *bestp*, and subsequent unlocking and looping continues until no subsequent improvement over the former *bestp* exists. (Usually this takes approximately five passes.) Vahid and Le extend the KL heuristic by replacing its cost function with a combined execution-time/area/communication metric, by redefining a move as a movement of a functional node across partitions (rather than a swap of nodes), and by reducing the running time of "next move selection" to be constant. Via these means, the authors are able to achieve nearly equal-quality partitions to simulated annealing in an order of magnitude less time. The running time of the algorithm is accelerated by consideration of task nodes at subroutine-level granularity.

5. PARTITIONING FOR RECONFIGURABLE SYSTEMS

The development of reconfigurable hardware and its realization in field programmable gate array (FPGA) technology has motivated continuing interest in hardware/software codesign platforms and frameworks. The evolution of FPGAs from small custom hardware to low-cost, powerful platforms able to store more than a million gates of logic has led to their adoption as coprocessor hardware. The flexibility and programmability of configurable devices allow them to be personalized to enhance the performance of a given application, or set of applications. Exciting research has investigated the potential performance boosts for real-time embedded systems, as well as general-purpose personal computer applications. A large number of processor/reconfigurable hardware architectures and design frameworks have been devised as a result.

The circumstances upon which a programmable device is reconfigured shape the potential performance gains realizable via its use, and mold the partitioning strategy that should be used to realize an effective co-design. In this work we differentiate between run-time reconfigurable (RTR) devices, whose functionality is changed during application execution, and semi-static reconfigurable devices. Unlike RTR devices, semi-static reconfigurable devices cannot be reprogrammed during an application's execution. However, a semi-static reconfigurable device might be modified between executions of different applications, or before subsequent executions of the same application, in order to provide the maximum hardware support for an application's current environment. (Note that this classification of device reconfiguration is potentially independent of the actual type of FPGA hardware used. Indeed, a single FPGA might be used as either a semi-static reconfigurable device or an RTR device, or each at different times.)

For semi-static reconfigurable devices, the flavor of the Hardware/Software Partitioning problem is not very different from the original formulation described previously. For each application execution, a different partitioning into hardware and software components might be configured onto the system. However, the decomposition of an application into software and reconfigurable hardware implementations could be devised and implemented using any of the aforementioned heuristics.

However, for architectures consisting of a processor and one or more fully dynamic RTR devices, the nature of the partitioning problem changes, as a spatial as well as temporal partitioning must be performed. The hardware/software partition must satisfy area constraints (the size of the FPGA configuration required at any given time) as well as performance constraints (where system execution time now includes the latency of every reconfiguration, as well as data/configuration communication and hardware execution time). Any violation of the hardware area constraint would lead to a subsequent reconfiguration, which might lead to a violation of system performance constraints. Therefore, the added dimension of time increases the complexity of partition design and evaluation.

Many reconfigurable architectures and design platforms leave the actual hardware/software partitioning choice to the system designer, or allow the designer to interactively explore the design space of partitioning options. Celoxica's DK-1 compiler interactively profiles the code and annotates the code with area and delay estimates, enabling the designer to make an informed partitioning decision [234]. The Napa C compiler, which targets National Semiconductor's NAPA1000 chip, uses user annotations in the application source code to guide system partitioning [219]. TOSCA's co-design evolution is driven by its Exploration Manager (EM) tool, which maintains a complete history of design alternatives explored by the user [235]. TOSCA's framework allows both direct intervention of the user in selecting a partition, and also an automatic partition selection guided by built-in evaluation criteria. Other design frameworks incorporating manual design partitioning include CASTLE [236] and POLIS [237].

PRISM (an acronym for Processor Reconfiguration Through Instruction Set Metamorphosis) was one of the earliest design environments seeking to augment the performance of a microprocessor via the synthesis of new operations [25]. The key innovation of this work was the choice to implement only instructions in the cooperating hardware logic. For instance, on a RISC-based system with a frequently executed subroutine *foo(a, b)*, the function *foo* could potentially become an architecture level instruction acting on two input registers (whose inner details are synthesized on the supporting hardware). Rather than begin with the ISA of the given processor, the PRISM compiler would initialize the application instruction set to be null. At compile time, the PRISM compiler would identify a set of operations that best represent the needs of the application. Any chosen operation not already supported by the microprocessor would be synthesized on the coprocessor. (As PRISM was a proof-of-concept implementation, the automation of this operation selection was never described.)

Another new system partitioning technique favored by reconfigurable architecture researchers is the utilization of accepted compiler techniques to find parallel regions of code, and then implement these regions in reconfigurable fabric. The search for reconfigurable candidate code begins with the identification of loop bodies that are frequently executed as well as computation-intensive. These loop bodies can be identified via static analysis or with a profiling tool. The rest of the partitioning involves pruning the list of candidate loops to satisfy system feasibility requirements and also maximize performance.

The Nimble compiler [238] extracts candidate loops as kernels for possible hardware implementation on RTR coprocessors, and applies hardware-oriented compiler optimizations (loop unrolling, fusion, pipelining) to generate multiple optimized versions of each loop. In determining which loops to synthesize, as well as which hardware implementation to use for each loop, the compiler is guided by a global cost function incorporating execution times on hardware and software as well as hardware reconfiguration time. The area constraints of the FPGA are considered, as violation would result in further reconfiguration. On Nimble's platform, only one loop can be synthesized in hardware at a time, and loops are executed purely sequentially, even if realized on the reconfigurable fabric.

DEFACTO, an end-to-end design environment, identifies loop candidates for realization in hardware via existing array data dependence analysis, privatization, and reduction recognition techniques [218]. Like Nimble, the DEFACTO compiler generates many specialized hardware versions of a given loop body. Working in tandem with an estimation tool, the design manager uses locality analysis to create a feasible partition that minimizes data communication and synchronization. DEFACTO specifically seeks to minimize the number of costly hardware reconfigurations, aiming to create hardware exhibiting semi-static behavior. Unlike Nimble, DEFACTO can produce a multi-way partitioning, supporting a single processor, multiple FPGA architecture.

Berkeley's BRASS project compiles to the Garp platform, which is composed of a MIPS microprocessor and a datapath-optimized reconfigurable coprocessor [221, 239]. Their compiler identifies candidate loops for reconfigurable acceleration. For each accelerated loop, they form a hyperblock containing the most frequently executed control paths of the loop body. This hyperblock will be the portion of the loop body that becomes implemented on the reconfigurable coprocessor. The predicated execution model of the hyperblock is realized in hardware via multiplexors, which act as the select instructions.

Markus Weinhardt and Wayne Luk of Imperial College also investigate loop synthesis techniques for reconfigurable coprocessors [240]. They refer to their approach as pipeline vectorization. Like the previous loop-identification approaches, compiler techniques are utilized to discover good candidates and subsequently resize them for high-performance FPGA implementation. Weinhardt and Luk only generate hardware from innermost loops with regular true loop-carried dependences (non loop-carried dependences, as well as anti- and output dependences, do not pose a threat to loop pipelining in hardware, and are thus not considered). Therefore, only loops whose current values are dependent upon previously written values (or in which no dependence on previous iterations exists) are considered. This dependence must be regular, meaning it spans the same distance (in iterations). A pipelined version of the loop, able to execute multiple iterations concurrently, is then synthesized on the hardware (containing circuit feedback loops to explicitly handle true loop-carried dependences). Realizing that this loop-investigation technique has limited applicability, Luk and Weinhardt apply classical loop transformations such as unrolling, tiling, and merging to reshape innermost loops into suitable hardware candidates. Therefore, the applicability of their technique is widened to innermost loops that would previously not have been considered, as they were too large or too small for hardware synthesis.

A recent approach for reconfigurable system partitioning [241] is based on the *Ant System (AS)* algorithm, a meta-heuristic optimization method inspired by the behaviors of ants. In the proposed algorithm, a collection of agents cooperate together to search for a good partitioning solution. Both global and local heuristics are combined in a stochastic decision making process in order to effectively and efficiently explore the search space. The approach is truly multi-way and can easily be extended to handle a variety of different system platforms.

The ant search algorithm, originally introduced by Dorigo et al. [242], is a cooperative heuristic searching algorithm inspired by the ethological study on the behavior of ants. It was observed that ants – who lack sophisticated vision – could manage to establish the optimal path between their colony and the food source.

A study [243] found that the ants use pheromone trails to communicate information amongst themselves. Though any single ant moves essentially at random, it will make a decision on its direction based on the "strength" of the pheromone on the paths that lie before it. As an ant traverses a path, it reinforces that path with its own pheromone. Therefore the "shortest" paths will maintain a higher amount of pheromone as opposed to the "longer" paths. A collective autocatalytic behavior emerges as more ants will choose the shortest trails (because the short trails have a higher amount of pheromone), which in turn creates an even larger amount of pheromone on those short trails. This means that those short trails will be even more likely to be chosen by future ants. The AS algorithm is inspired by this observation. It is a population based approach where a collection of agents cooperate together to explore the search space. They communicate via a mechanism imitating the pheromone trails.

The ant search formulation for the partitioning uses an augmented task graph, which serves as a generic mathematic model for the multi-way system partitioning problem. An *augmented task graph (ATG) $G' = (T,E',R)$* is an extension of the task graph G. It is derived from G as follows; given a task graph $G = (T,E)$ and a system architecture R, each node $t_i \in T$ is duplicated in G'. For each edge $e_{ij} = (t_i, t_j) \in E$, there exist r directed edges from t_i to t_j in G', each corresponding to a resource in R. Our algorithm uses these augmented edges to make a local decision at task node t_i about the binding of the resource on task t_j. We call this an augmented edge. An augmented task graph for a platform with 3 programmable components is shown in Figure 29. The right graph of the figure is a possible solution to the partitioning problem using the ATG.

To make the model complete, a dot operation is defined, which is a bivariate function $f_{ik} = t_i \bullet r_k, \forall t_i \in T, \forall r_k \in R$. It provides local cost estimation for assigning task t_i to resource r_k. Assuming we are only concerned with the execution time and hardware area in our partitioning, we can let f_{ik} be a two item tuple, i.e. $f_{ik} = t_i \bullet r_k = \{time_{ik}, area_{ik}\}$. Obviously, other items, such as power consumption estimation, can be easily added if they are considered. The dot operation can be viewed as an abstraction of the work performed by the cost estimator. This allows our algorithm to customize towards the system platform that we are targeting. Each platform will have different estimation functions depending on the characteristics of that platform.

Currently, we have implemented several very simple cost estimators for system components like a DSP processor, RISC processor and FPGA. Future research for more accurate component estimation models is needed. Specifically, we must study the power/performance tradeoffs that we can make when partitioning the application onto the device.

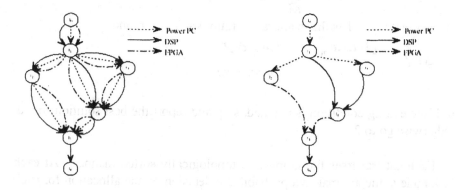

Figure 29 The augmented task graph (ATG) for 3 way partitioning (PowerPC, DSP, FPGA) and a particular solution for the 3 way partitioning.

The AS algorithm for multi-way system partitioning is a multi-agent stochastic decision making process that combines local and global heuristics during the search process. Each agent (ant) traverses the ATG and attempts to create a feasible partitioning by selecting the next move probabilistically according to the combined heuristics. The quality of each solution is measured using the system objective function with consideration to the system constraints. The quality of the solution (as dictated by the objective function) is used to reinforce good solutions. The global heuristic information is distributed as pheromone trails on the edges of the ATG.

The proposed algorithm proceeds as follows:

1. Initially, associate each augmented edge e_{ijk}, in the ATG with a pheromone τ_{ijk}; the value of the pheromone on each augmented edge is initially set at the value τ_0.

2. Put m ants on task node t_0;

3. Each ant crawls over the ATG to create a feasible partitioning $P^{(l)}$, where l = $1, \dots ,m$;

4. Evaluate the partitions generated by each of the m ants. The quality of the particular partition $P^{(l)}$ is measured by the overall execution time.

5. Updated the pheromone trails on the edges as follows:

$$\tau_{ijk} \longleftarrow (1 - \rho)\tau_{ijk} + \sum_{l=1}^{m} \Delta \tau_{ijk}^{(l)}$$

where $0 < \rho < 1$ is the evaporation ratio, $k = 1, ..., r$, and

$$\Delta \tau_{ijk}^{(l)} = \begin{cases} Q / time_{P(l)} & if\ e_{ijk}' \in P^{(l)} \\ 0 & otherwise \end{cases}$$

6. If the ending condition is reached, stop and report the best solution found. Otherwise go to 2.

Each ant traverses the graph in a topologically sorted manner. At each task node t_i, the ant makes a probabilistic decision on the allocation for each of its successor task nodes t_j based on the pheromone on the edge. The pheromone is manipulated by the distributed global heuristic τ_{ijk} and a local heuristic for assigning the resources to different tasks. An ant at t_i chooses that resource r_k is assigned to node t_j according to the probability:

$$p_{ijk} \quad \frac{\tau_{ijk}^{\alpha} \eta_{jk}^{\beta}}{\sum_{l=1}^{r} \tau_{ijl}^{\alpha} \eta_{jl}^{\beta}}$$

Here η_{jk} is the local heuristic if t_j is assigned to resource r_k i.e. $t_i \bullet r_k$. Currently, this local heuristic is based on a simple weighted combination of task graph latency and cost of the system. We need to study advanced techniques for these local estimation of techniques. In particular, we should take the additional communication costs of transmitting data between components. Furthermore, we should exploit the dynamic reconfiguration opportunities. Upon entering a new node t_j, the ant must decide upon the allocation of this node t_j based upon the choices of the immediate predecessors of t_j. In our algorithm, this decision is currently made probabilistically based on the distribution of the predecessors guesses.

At the end of each iteration, the pheromones on the edges are updated according to Step 4. First, a certain amount of pheromone is evaporated. Secondly, the "good" edges, which correspond to the best partition solutions found in the previous iteration, are reinforced. This reinforcement creates additional pheromone on the edges that are included on partition solutions that provide shortest execution time for the task graph. Alternative reinforcement methods can be applied here. In each run of the algorithm, multiple iterations of the above steps are conducted. Two ending possible stopping conditions are: 1) the algorithm ends after a fix number of iterations, or 2) the algorithm ends when there is no improvement found after a number of iterations.

To test the effectiveness of the algorithm, a task level benchmark suite based on the MediaBench applications [203] was created. Each testing example is formed via a two step process that combines a randomly generated DAG with multimedia software function from the MediaBench applications. The testing suite contains 25 task graphs.

The AS algorithm is compared with brute force search method and simulated annealing. It is observed in our experiments that the AS algorithm quickly and effectively converges to a solution. Figure 30 shows the cumulative distribution of the number of solutions found by the AS algorithm plotted against the quality of those solutions. The x-axis gives the solution quality compared to the overall number of solutions. The y-axis gives the total number of solutions (in percentage) that are worse than the solution quality. For example, looking at the x-axis value of 2%, less than 10% of the solutions that the AS algorithm found were outside of the top 2% of the overall number of solutions. In other words, over 90% of the solutions found by the AS algorithm are within 2% of all possible partitionings. The number of solutions drops quickly showing that the AS algorithm finds very good solutions in almost every run. In our experiments, 2, 163 (or 86%) solutions found by AS algorithm are within the top 0.1% range. Totally 2, 203 solutions, or 88.12% of all the solutions, are within the top 1% range. The figure indicates that a majority of the results are qualitatively close to the optimal. We compared the results of AS with the simulated annealing method for system partitioning. The SA implementation is similar to that of Wiangtong et al. [244]. Figure 30 also shows the results of several implementations of simulated annealing. Here SA-50 has roughly the same execution time of ant search, while SA-500 (SA-1000) takes approximately 10 (20) times longer than our ant search algorithm. We can see that with substantially less running time, ant search achieves better results than the SA approach, even when compared with a much more exhaustive SA session like SA-1000.

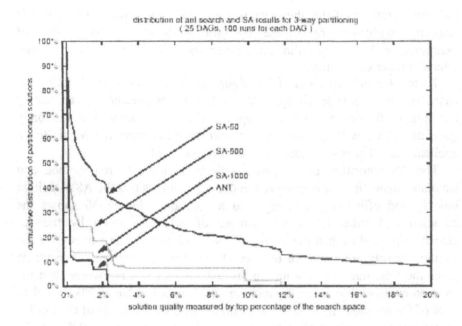

Figure 30 Comparing the solution quality of Ant Search with Simulated Annealing.

The preliminary results for the ant search algorithm are quite promising. The algorithm consistently provides near optimal partitioning results with very minor computational cost. However, there is much work that needs to be done. The task graph application model used in preliminary experiments lacks many of the features necessary for system partitioning. This requires that we extend the algorithm to handle loops (currently we can only handle acyclic graphs). Furthermore, hybrid methods for system partitioning look promising. The ant system algorithm is effective because it is tightly coupled to the problem, unlike search space exploration algorithms like simulated annealing and Tabu search. However, these other algorithms have their benefits. We believe that hybrid algorithms that use different algorithms (e.g. first use AS, then do a low temperature annealing) have much promise, especially when we use more complex estimation functions for targeting real programmable devices. Furthermore, hierarchical partitioning looks like a good way to perform combined task and instruction scheduling. Hierarchical partitioning will provide less reliance on instruction level estimation models, which will allow us to better explore the mapping of the application to the programmable device.

6. SUMMARY

Hardware/software partitioning is a fundamental task in system design. Good partitioning algorithms are needed to assign computing tasks of an application specification to the resources of the computing device. In this chapter, we looked at different algorithms and methodologies for solving the hardware/software partitioning problem.

We started by looking at the general partitioning problem. It has been studied quite extensively in design automation. We looked at the general problem and described its relationship with the hardware/software partitioning problem. Then we discuss formulations and models used to describe the hardware/software partitioning. We conclude the chapter by surveying a number of different algorithms used to solve the problem.

Chapter 7

INSTRUCTION GENERATION

1. INTRODUCTION

Computational devices are becoming more complex. The number of transistors on a single die – dictated by Moore's Law – is increasing exponentially. This allows a *system-on-chip* – a variety of different computing devices interacting as a complete system on a single die. An additional benefit of Moore's Law is the decreasing cost of computations leading to a ubiquity of embedded systems. Much like a system-on-chip, these embedded systems are a complex interaction of many different computational devices, embedded within a larger entity e.g. a car, telephone, building, etc.

With the proliferation of computing systems comes a need for specialization; each use of the system is tailored for a specific set of applications. For example, a digital system embedded within a cellular phone will encounter DSP-type applications. It most likely needs to perform operations like analog to digital conversions (and vice-versa), FFT, and filtering. Therefore, if we customize the embedded system to such operations, we gain increased performance, power/energy reduction and smaller silicon footprint. This tends towards the use of ASICs for such systems.

Yet, an ASIC is extremely inflexible; once the device is fabricated, the functionality cannot be changed. For example, if a new cellular communication standard emerges, we must throw away our phone and buy a new one customized for the new standard. However, if a new standard appears and the embedded system is flexible, we can change the functionality of the system to handle the migration from one standard to another. Another increasingly important trend is time to market. The initial market share of a product that is released first is inherently larger than that of a product that is released much later. In fact, we are seeing that time to market is becoming vital to the success of the product. The public accepts new products at a fanatical pace. Products once took 10+ years to gain consumer acceptance (e.g. television, radio). Now, products permeate the market in under a year (e.g. DVD, MP3 players). Both of these trends accentuate the need for flexible devices like a general purpose processor. The more applications/standards that the device can service, the more companies will use it, as they want to reach the market quickly and service as many markets as possible.

We have conflicting forces pushing digital systems in two seemingly separate directions. On one hand, computing systems must be specialized to meet the performance, power, energy, and area constraints. In this regard, ASICs are the answer. On the other hand, time to market and flexibility constraints push for general purpose systems. Obviously, there is a tradeoff between the flexibility (general purpose) and performance (application specific) of the system. We need methods to allow us to perform the tradeoffs between these two metrics. We must be able to customize a system towards the tasks that it will most likely perform and give it the flexibility to adapt over the course of its lifetime.

Customized instructions are one way to explore the tradeoff between customization and flexibility. They approach the problem through the customization of a general purpose device. If we know the *context* – a set of applications that will likely run on a system, we can look for commonly occurring, computational sequences within the various applications of the context. These customized instructions can be optimized for high-performance, low energy/power consumption and/or small area. For example, it is well known that the computational sequence, multiply-accumulate (MAC), occurs frequently in DSP applications. The MAC unit would be an ideal candidate for a customized instruction when the context is DSP applications. In this chapter, we develop a systematic method for customized instruction generation and explore the theoretic aspects of *instruction generation* – the process of finding commonly occurring computational patterns within a context.

Customized instructions can serve to optimize two general purpose devices – the processor and the FPGA. *Application specific instruction set processors (ASIPs)* take a small processor core and add customized instructions to service the specific context of the applications. The PICO project [245] aims to automatically generate the customized instructions based on a specific application. They use a VLIW core and generate nonprogrammable hardware accelerators (NPA), which are akin to systolic arrays. The interface between the NPAs and the processor core is automatically synthesized. The user is responsible for identifying the customized instructions in the form of loop nests. Another product, Tensilica's Xtensa processor [246], takes the customized instructions as an input in the form of a *hardware description language (HDL)* called the Tensilica Instruction Extension (TIE) Language. It incorporates the instruction into a compiler allowing the user to execute the instruction through an intrinsic function. Our instruction generation algorithms can serve the Xtensa and PICO frameworks to automatically find the customized instructions.

The FPGA is another general purpose computing device, albeit quite different from a general purpose processor. The FPGA has the benefit of adapting its architecture directly to the application that it implements. Data intensive applications running on an FPGA can achieve up to 100x increased performance as compared to the same application running on a processor [247-250]. This mainly comes from the ability to customize at the architecture level. The architecture of an application running on an FPGA is completely flexible, whereas the processor architecture is fixed.

Yet, the architecture of an FPGA is still tailored for the general case. Adding *macros* to the architecture could customize the FPGA. Macros are hard or soft reconfigurable computational sequences. A *hard macro* is a fixed ASIC core embedded into the fabric of the FPGA. The embedded multipliers of the Xilinx Virtex series are an example of a hard macro. A *soft reconfigurable macro* is a sequence of computations that are implemented as a fixed entity on the FPGA fabric. Examples of soft reconfigurable macros are the components of the Xilinx CoreGen library.

In this sense, reconfigurable architectures are moving away from reconfiguration exclusively at the gate level and moving towards a *hybrid reconfigurable architecture*. Hybrid reconfigurable architectures contain reconfigurability at multiple levels of the *computational hierarchy* (see Figure 31). The computational hierarchy is the level of abstraction that computations may be implemented. One level of the computational hierarchy is the gate or Boolean level. At this level, every computation is built up from the Boolean (gate) level computations. The FPGA is an example of a device that functions at this level. A device can also be reconfigured at the microarchitecture or architecture level. These levels of computational hierarchy have coarser basic computational units. A basic unit at the microarchitecture level is on the level of an arithmetic function. PipeRench [251] and RaPiD [252] are examples of this type of reconfigurable system. At the architecture level, the basic unit of computation is more coarse grained, for example the RAW project [253].

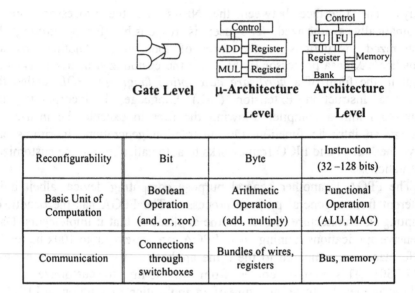

Reconfigurability	Bit	Byte	Instruction (32 –128 bits)
Basic Unit of Computation	Boolean Operation (and, or, xor)	Arithmetic Operation (add, multiply)	Functional Operation (ALU, MAC)
Communication	Connections through switchboxes	Bundles of wires, registers	Bus, memory

Figure 31 A comparison between three levels of the computational hierarchy. The gate level is the most flexible with the architecture level being the least flexible but the architecture level has the fastest reconfiguration time. The performance, area and power/energy consumption depend on the type of operation being implemented.

Reconfigurability at the various levels of the computational hierarchy gives many tradeoffs in terms of flexibility, reconfiguration time, performance, area and power/energy consumption. A fine-grained reconfigurable device (gate level) is extremely flexible; it can implement any application. But, the flexibility comes at a cost. The routing architecture must allow a connection from any part of the chip to any other part of the chip. Switchboxes are used to enable this sort of flexibility. The switchboxes are composed of many transistors to enable a flexible routing. Compared to a direct connection, it is apparent that switch boxes add much overhead to the area, delay (performance), and power/energy consumption. Furthermore, the implementation of an arithmetic unit (e.g. an adder) on a fine-grained reconfigurable device consists of programming each gate of the arithmetic unit. Since the gates are designed to be extremely flexible – you can implement any Boolean function on any gate – the arithmetic unit will be large, slow and power/energy hungry as opposed to the same arithmetic unit that is designed specifically for implementing that function, as is the case for a device that is reconfigurable at the microarchitecture level. On the other hand, the area, delay, power/energy consumption of implementing a Boolean function favors a gate level reconfigurable device. If we implement a one bit "and" function on a device that is reconfigurable at the architectural level, the function will be over designed. It will implement an "and" instruction and 31, 63 or 127 "and" operations will be unnecessarily performed, depending on the size of the instruction. As you can see, there are benefits for reconfigurability at various levels of the computational hierarchy. A hybrid reconfigurable architecture allows us to mix and match these levels to tailor to the applications at hand.

Examples of hybrid reconfigurable systems include Garp [254], which couples a MIPS-II processor with a fine-grained FPGA coprocessor on the same die. The *Strategically Programmable System (SPS)* architecture [255] combines memory blocks, *Versatile Programmable Blocks (VPBs)* – embedded ASIC blocks that perform complex instructions – into a LUT-based fabric. Many other academic projects can be called a hybrid reconfigurable system, for example Dynamically Programmable Gate Array (DPGA) [256] and Chimaera [257].

In addition, several industrial projects fall into the category of hybrid reconfigurable systems. One example is the Virtex-II devices from the new Xilinx Platform FPGAs, which embed high-speed multipliers into their traditional LUT-based FPGAs. Also, the CS2112 Reconfigurable Communications Processor (RCP) from Chameleon Systems, Inc. contains reconfigurable fabric organized in slices, each of which can be independently reconfigured.

To understand how instruction generation works in hybrid reconfigurable systems, we consider the SPS project. SPS consists of VPBs embedded into a LUT-based fabric. It is targeted towards a specific context. An example of a specific SPS architecture is shown in Figure 32. Because the VPBs are hard macros, implementing an operation on them as opposed to on the reconfigurable fabric will give lower power and energy consumption. Additionally, the VPBs require no time to program. Hence, the SPS architecture can be reconfigured faster than an FPGA. Furthermore, the performance of the operation on the VPB will be much better than the performance of the same operation on the reconfigurable fabric. But, we must carefully consider the type of functionality for the VPB. If the applications of the context never use the VPBs, then they are wasting space on the chip.

A designer of an SPS architecture can specify the functionality of the VPBs towards the targeted context. The designer must consider the types of operations that make up the applications of the context. The operations must occur frequently. Furthermore, the operations must give performance or other benefits (e.g. reduced power consumption) when implemented as VPB as opposed to on the reconfigurable fabric. There are many tradeoffs to consider when choosing the functionality of the VPB. A tool to determine these tradeoffs would help the designer pick the functionality. Ideally, the tool would automatically determine the functionality of the VPBs based on the given context. This is exactly the problem of instruction generation.

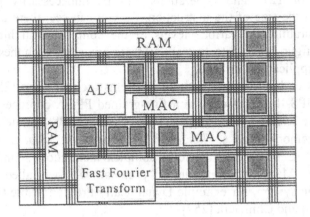

Figure 32 Example of the Strategically Programmable System (SPS) – a hybrid reconfigurable system. Functional units are embedded within a reconfigurable fabric. This SPS architecture would be specific to the DSP context.

Instruction generation is also useful for the generation of soft reconfigurable macros. Soft reconfigurable macros act as a "black box" to a designer. The input, output and functionality of the soft reconfigurable macro are given to the designer, but the actual implementation of the macro on any specific reconfigurable architecture is abstracted away. In this sense, the terms soft reconfigurable macro and soft reconfigurable IP (intellectual property) are synonymous. A soft reconfigurable macro is customized for each reconfigurable architecture. For example, the Xilinx CoreGen library has IPs like decoders, filters, memories, etc. Each of these functions is customized for the architecture on which it runs. The CoreGen components are generated based on what Xilinx believes their customers will use. Also, the functionalities of the CoreGen IPs are well-known entities. Instruction generation is useful to find unknown, irregular computational patterns that are not immediately apparent from looking at the code of the applications in a context.

Soft reconfigurable macros give many benefits. They allow the designer to work at a higher level of abstraction. Instead of dealing with basic arithmetic operations like addition, multiplication, etc, the designer can implement the application using function or block level structures. Matlab works at this level and is extremely popular in the signal processing community. Additionally, the soft reconfigurable macros can be highly optimized. A person familiar with the underlying architecture can design each macro. Therefore, the macros will be more efficient than if someone who doesn't understand the underlying architecture implemented them or if they were designed from basic operations using the synthesis flow. Finally, the compilation time of the applications using the macros is reduced. The macros can be pre-placed and routed. Therefore, the tool or designer must only determine the location of macro on the reconfigurable fabric. The placement and routing of the macro – an extremely time consuming task – is unnecessary.

In summary, instruction generation is an extremely important concept for context-specific architectures. Whether the underlying architecture is derived from a general purpose processor or a reconfigurable architecture, instruction generation is essential to the flexibility and performance of the system. Furthermore, instruction generation in the form of soft reconfigurable macros can reduce the time for synthesizing an application to reconfigurable architectures.

In the next section, we look at the role of instruction generation in a reconfigurable system compiler. The following section formalizes the problem of instruction generation. We discuss the specifics of the formulation of the instruction generation problem in Section 3. Section 4 discusses basic background knowledge pertaining to instruction generation. Section 5 discusses our initial work on template generation, an iterative, constructive algorithm for generation of sequential templates using graph profiling and edge contraction. In Section 6, we discuss the latest work in template generation: an algorithm that uses slack calculations on dataflow dependence to generate parallel templates. In Section 7, we present experimental results. Then, in Section 8, we discuss related work. We conclude in Section 9.

2.   INTEGRATION OF INSTRUCTION GENERATION WITH RECONFIGURABLE SYSTEM SYNTHESIS

Reconfigurable system synthesis has two different uses. One flow is architecture generation and another flow for application mapping. Figure 33 gives a high level overview for both of these flows. We use the SPS co-compiler as a representative reconfigurable system compiler.

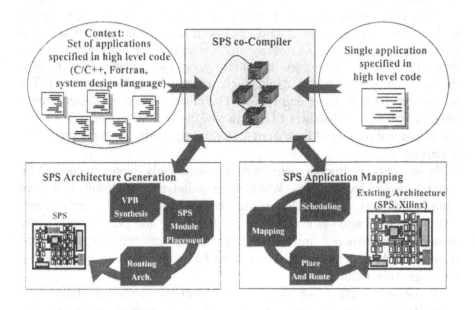

Figure 33 Two flows for the SPS system. The flow on the left is the SPS architecture generation flow. The application mapping flow is on the right. The co-compiler interfaces with both flows.

In both flows, the first task is to translate each of the applications of the context or the single application into a form that is suitable to interface with architecture generation and application mapping, respectively. The applications are given in a high level language. For example, the applications could be written in C/C++, Fortran or some other system design language, like SystemC [153], SpecC [151] or Esterel [258]. We must choose an intermediate representation (IR) for the co-compiler. We use the *control data flow graph (CDFG)* as the intermediate representation. A CDFG is a directed, labeled graph with data nodes corresponding to operations (addition, multiplication, shift) and control nodes that dictate the flow of the program. Control nodes allow branches and loops. The edges between the nodes represent control and data dependencies. The CDFG offers several advantages over other models of computation. The techniques of data flow analysis (e.g. reaching definitions, live variables, constant propagation, etc.) can be applied directly to CDFGs. Also, many high-level programming languages (e.g. Fortran, C/C++) can be compiled into CDFGs with slight modifications to pre-existing compilers; a pass converting a typical high-level IR into control flow graphs and subsequently CDFGs is possible with minimal modification. Therefore, we can leverage the front-end of many existing compilers for our reconfigurable system compiler.

On the other hand, CDFGs only have the ability to describe instruction level parallelism. In order to specify a higher level of parallelism, another model of computation (MOC) must be used. But, we could embed a CDFG into another MOC – one that can describe a higher level of parallelism. For example, we could embed CDFGs into finite state machines (FSM). Lee's *charts [259] do something similar; they embed synchronous data flow graphs into a FSM.

Instruction generation has a role in both the architecture generation flow and the application mapping flow. In the architecture generation flow, instruction generation determines the functionality of the hard macros – the VPBs in SPS. During application mapping, instruction generation is needed to determine soft reconfigurable macros.

There are many other tasks in the architecture generation and application mapping flows. For example, we must determine the exact placement of the VPBs, their interface with the reconfigurable fabric, the routing architecture, the exact number of each type of VPB and so on. We focus on generating the VPB functionality and refer the interested reader to our other papers on these subjects [255, 260, 261].

3. PROBLEM FORMULATION

Instruction generation is analogous to the problem of regularity extraction, which has been used to detect recurring subcircuits in the logic synthesis and physical design stages of a CAD flow. Template Generation refers to the process of selecting a set of candidate IPs from an application represented as a CDFG. Vertices in a CDFG represent basic blocks, which occur in a program; edges represent (conditional or unconditional) branching instructions. A basic block is a set of consecutive statements in a program such that none of the statements in the block is either a branch statement itself or the target of a branch statement. Once a basic block is entered, an exit is guaranteed. Each vertex of a CDFG can be thought of as containing its own dataflow graph (DFG). In a DFG, vertices represent inputs, outputs, or arithmetic operations, such as addition or multiplication; edges represent data dependencies between inputs, outputs, and operations. An example of a small function and its decomposition into a control-flow graph (CFG) is presented in Figure 34.

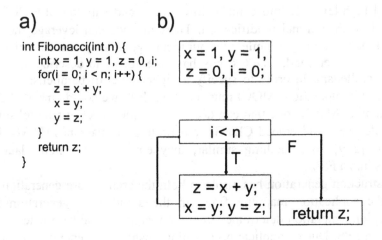

a)

```
int Fibonacci(int n) {
    int x = 1, y = 1, z = 0, i;
    for(i = 0; i < n; i++) {
        z = x + y;
        x = y;
        y = z;
    }
    return z;
}
```

b)

x = 1, y = 1,
z = 0, i = 0;

i < n   F

T

z = x + y;
x = y; y = z;   return z;

Figure 34 A small C Function (a) and its CFG Representation (b)

Templates are repeated occurrences of (possibly interdependent) nodes and edges in a DFG. Each vertex in a DFG represents some simple operation, such as addition or multiplication. Templates can be thought of as vertex clusters, or super-nodes, which represent instructions at the architecture level. In this paper, we define two different types of templates. Sequential templates correspond to nodes connected by directed edges in the DFG (i.e. nodes which represent a sequential execution operation). Parallel templates correspond to nodes whose operations can be scheduled for simultaneous execution. Sequential templates contain direct data dependencies (as they follow DFG edges) and thus cannot directly increase inherent parallelism in the execution. Parallel templates have no data dependencies between them, but may restrict the scheduling mechanism at lower synthesis stages. An example of a DFG with sequential and parallel templates is shown in Figure 35.

Template generation describes the process of finding a suitable set of templates given a DFG or set of DFGs (from a CDFG). An algorithm first described in [262] is suitable for the generation of sequential templates. A newer algorithm for the generation of parallel templates, first published in [263], is the primary contribution of the work described herein.

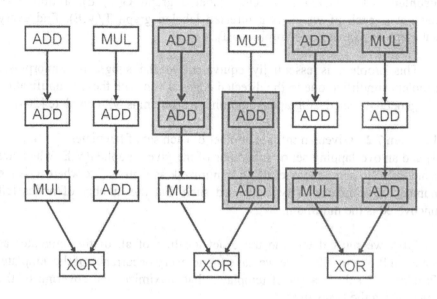

Figure 35 A simple DFG (a) with sequential (b) and parallel (c) templates.

3.1   Template Matching

Regularity refers to the repeated occurrence of computational patterns e.g. multiply-add patterns in an FIR filter and bi-quads in a cascade-form IIR filter. A *template* refers to an instance of a regular computational pattern.

We model an algorithm, circuit or system using a digraph $G(V,E)$. The nodes of the graph correspond to an instance of basic computational units. Examples of node types are add, multiply, subtract, etc. Each node has a label consistent with the type of operation that it performs. The edges of a graph model the dependencies between two operations. For instruction generation, the graph under consideration is a data flow graph.

It should be noted that systems/circuits must often be modeled by hypergraphs. A *hypergraph* is like an ordinary graph, but each *hyperedge*, connects multiple vertices instead of two as in a normal digraph. Extending our algorithms to consider hypergraphs is fairly straightforward.

We consider labeled digraphs in this work as we mainly target instruction generation, which use compiler data flow graphs; data flow graphs can be sufficiently modeled using labeled digraphs.

There are two general problems associated with template matching:

Problem 7.1:   Given a directed, labeled graph G(V, E), a library of templates, each of which is a directed labeled graph $T_i(V,E)$, find every subgraph of G that is isomorphic to T_i.

This problem is essentially equivalent to the subgraph isomorphism problem simplified due to the directed edges. Even with these simplification the general directed subgraph isomorphism problem is NP-complete [264].

Problem 7.2:   Given an infinite number of each set of templates $\Omega = T_1, \ldots ,$ T_k and an overlapping set of subgraphs of the given graph G(V,E) which are isomorphic to some member of Ω; minimize k as well as $\Sigma\ x_i$ where x_i is the number of templates of type T_i used such that the number of nodes left uncovered is the minimum.

First, we must determine the exact location of all of the templates as stated in Problem 7.1. Once we have found every occurrence of the template, Problem 7.2 selects a set of templates that maximizes the covering of the graph using the templates.

We want to minimize both the number of distinct templates that are used in the covering while minimizing the number of instances of each template. Additionally, we want to cover as many nodes as possible. This problem is a fusion of the graph covering and the coin changing problems. It differs from the graph covering as it allows multiple instances of template in its covering. The coin changing problem tries to find the minimum number of coins to produce exact change; this is similar to minimizing the number and types of templates to cover the graph.

The classical compiler problem of instruction selection falls into the realm of template matching. We are given the templates or instruction mappings corresponding to a directed, labeled graph of the program more commonly known as the IR. Instruction selection is directly related to Problem 7.2, with possibly additional objectives e.g. minimize runtime of the code, size of the code, etc.

3.2   Template Generation

Up until this point, it was assumed that the templates were given as an input. However, this may not always be the case; an automatic regularity extraction algorithm must develop its own templates.

Consider instruction generation for hybrid reconfigurable architectures. The instructions (templates) for traditional processors are fixed according to the target architecture. Since we are dealing with hybrid reconfigurable architectures, the instructions are not fixed. It is possible to arrange the reconfigurable fabric to perform virtually any combination of basic operations. Therefore, the instruction templates are not fixed in reconfigurable architectures and template generation is a necessary step for the hybrid reconfigurable architecture generation and compilation.

The configurable fabric allows the designer to implement custom instructions as well as perform any fixed instructions on more traditional embedded processing units. In essence, the compiler must perform instruction generation and selection, equivalently template generation and matching.

Additionally, template generation is useful for creating macro libraries for both ASIC and FPGA architectures [265]. Also, template generation is needed for effective system-level partitioning [266].

4. PRELIMINARY DISCUSSION AND BACKGROUND MATERIAL

4.1 Dataflow Model

A dataflow graph (DFG) $G = (V, E)$ is used to describe computation within basic blocks of a program. A DFG is modeled as a directed acyclic graph (DAG), where each vertex represents a simple machine operation and each edge represents a dependency between two sequentially ordered operations. Each vertex v maintains a list of incoming neighbors in_v and out_v; in-degree$_v$ and out-degree$_v$ are defined as the number of incoming and outgoing neighbors of v respectively.

For the purposes of this model, we assume that the set of possible vertex operations is given. In practice, the compiler designer will be responsible for determining the exact set of operations. Let $Op = \{op_1, ..., op_m\}$ be the set of operations. For the purposes of simplification, we map each operation to a set of integer types $T = \{0, ..., m-1\}$ where op_i maps to value $i-1 \in T$. The operation represented by each vertex is an inherent property of the vertex in this model. Thus, each vertex v has a type $T_v \in T$. Types for edges are defined similarly. Edge $e = (u, v)$ has a type $T_e = (T_u, T_v) \in T \times T$.

A well-known result from graph theory is that every DAG has at least one source and at least one sink. In the dataflow model, the source vertices represent inputs to a dataflow graph and the sink vertices represent the outputs. In our model, inputs to vertices may be either variable (VAR) or immediate (IMM). Variables are stored in either register or memory, and immediates are constant values, which are simply wires on the circuit connected to power and ground. Outputs are stored either in registers or written to memory.

4.2   Templates and Clustering

In this section, we extend our dataflow model to facilitate templates. Templates are individual vertices that can replace a subset of vertices (and possibly a subset of edges) in a dataflow graph (DFG). The process of replacing a subset of vertices (and possibly edges) by a template vertex is called clustering; reverse clustering is the process of removing a template from a DFG and replacing it with the vertices (and possibly edges) that it encapsulates. Clustering and reverse clustering are inverse operations.

In [262], we proposed instruction generation as an optimization based on a compiler's representation of a program: a CDFG. Later stages of a design flow may wish to undo a template, for the purpose of optimizing some criteria other than regularity. A software compiler cannot be aware of the detailed hardware implementation of the instructions (also referred to as templates) it generates.

Another question to be addressed is whether templates can subsume other templates, creating a hierarchy of templates. We decided against storing templates hierarchically because we believe it would complicate our test for template equivalence. Instead, we apply a reverse-cluster operation to each template that is subsumed.

In order to allow for clustering and reverse clustering, each template T must be augmented with several fields: internal vertices, internal edges, internal incoming edges, and internal outgoing edges. None of these edges should be confused with T's incoming and outgoing edges, which connect T to other vertices in the DFG.

Let $G = (V, E)$ be a dataflow graph. Suppose $G' = (V', E')$, $V' \subseteq V$, $E' \subseteq E$, is a subgraph of G whose vertices will be clustered by a template t. Let $G^T = (V^T, E^T)$ be the resulting graph after V' is clustered. Define the following attributes of t:

Internal dag: the dag $G' = (V', E')$ (subgraph subsumed by t).
Internal vertices: the set of vertices V'
Internal edges: the set of edges $E' = \{(u, v) \mid u \in V', v \in V'\}$
Internal incoming edges: the set of edges $E'_{in} = \{(u, v) \mid u \in V - V', v \in V'\}$
Internal outgoing edges: the set of edges $E'_{out} = \{(u, v) \mid u \in V', v \in V - V'\}$

The clustering and reverse clustering operations are presented as algorithms in Figure 36 and Figure 37 respectively.

Theorem 7.1: The Clustering Operation has time complexity $O(|V| + |E|)$.

Proof: The Clustering Operation in Figure 36 consists of four loops, none of which are nested. In the context of this algorithm, set union is to add an element to a set, which can be accomplished in constant time. The bounds of the four loops are V, E, E'_{in}, and E'_{out} respectively. Since $E'_{in} \subset E$ and $E'_{out} \subset E$, the time complexity of the respective loops are bounded by $|E|$. Thus, the total time complexity is $O(|V| + |E|)$. ❑

Theorem 7.2: The Reverse Clustering Operation has time complexity $O(|V| + |E|)$.

Proof: Consider the Reverse Clustering Operation in Figure 37. First, consider the sets: $e^t_{in} = \{(u, t)\} \subset E^T$ and $e^t_{out} = \{(t, v)\} \subset E^T$. Since G^T is a DAG, $e^t_{in} \cap e^t_{out} = \Phi$; otherwise, there would be a self-loop (t, t). From the construction of t in the Cluster operation, $|e^t_{in}| \leq |t.\text{internal\_incoming\_edges}|$ and $|e^t_{out}| \leq |t.\text{internal\_outgoing\_edges}|$. Additionally, note that $E \cap E^T = E^T - e^t_{in} - e^t_{out}$. Therefore, $|E^T| \leq |E|$, so $E^T = O(|E|)$. Similarily, $V^T = O(V)$ since $|V^T| = |V| - |t.\text{internal\_vertices}| + 1$ and $|t.\text{internal\_vertices}| \geq 2$ for all templates t.

The for-loop examines all edges in E^T, requiring $O(|E|)$ time. Statement 2 entails the construction of V from VT, requiring $O(|V|)$ time. Since t.internal_edges $\subset E$, statement 3 requires $O(|E|)$ time. From the argument above, statement 4 requires $O(|E|)$ time as well. Statements 5, 6, and 7 are trivial operations. Thus, the total time complexity is $O(|V| + |E|)$. ❑

Algorithm: Cluster(G, V')
Inputs: DFG G = (V, E), Subset of vertices to be clustered V' ⊆ V
Output: DFG with template $G^T = (V^T, E^T)$
Variables: t – the template vertex
                  E', E'$_{in}$, E'$_{out}$ – sets of edges.
                  V" – set of vertices

E' = , E'$_{in}$ = E'$_{out}$ = NULL;   V" = NULL
For each vertex v ∈ V'
If v is a template
V" = V" ∪ v.internal_vertices
G = Reverse_Cluster(G, v)
For each edge e = (u, v) ∈ E
If u ∈ V' and v ∈ V'
E' = E' ∪ e
Else if u ∈ V and v ∈ V'
E'$_{in}$ = E'$_{in}$ ∪ e
Else if u ∈ V' and v ∈ V
E'$_{out}$ = E'$_{out}$ ∪ e
V' = V' ∪ V"
t.internal_vertices = V'
t.internal_edges = E'
t.internal_dag = G' = (V', E')
t.internal_incoming_edges = E'$_{in}$
t.internal_outgoing_edges = E'$_{out}$
V = (V – V') ∪ {t}
E = E – E' – E'$_{in}$ – E'$_{out}$
For each edge e' = (u, v) ∈ E'$_{in}$
E = E ∪ {(u, t)}
For each edge e' = (u, v) ∈ E'$_{out}$
E = E ∪ {(t, v)}
$V^T = V$
$E^T = E$
return $G^T = (V^T, E^T)$

Figure 36 The Clustering Algorithm

Algorithm: Reverse_Cluster(G^T, t)
Inputs: DFG with template $G^T = (V^T, E^T)$, Template t ∈ V
Output: DFG without template t, G = (V, E)

For each edge e ∈ {(u, v) ∈ E^T | u = t or v = t}
$E^T = E^T - \{e\}$
$V^T = V^T \cup$ t.internal_vertices − {t}
$E^T = E^T \cup$ t.internal_edges
$E^T = E^T \cup$ t.internal_incoming_edges $\cup$ t.internal_outgoing edges
$V = V^T$
$E = E^T$
return G = (V, E)

Figure 37 The Reverse Clustering Operation

4.3   Template Type Equivalence: Isomorphism

Every vertex v in a DFG corresponds to some basic machine operation, for example addition or multiplication. v has a type t(v) corresponding to v's operation. To determine whether two vertices implement the same operations, we simply compare their types. Types for directed and undirected edges are defined follows. If e = (u, v) is directed, then t(e) = (t(u), t(v)). Otherwise, if e = (u, v) is undirected, then t(e) = (min(t(u), t(v)), max(t(u), t(v)).

We desire a similar method for comparison between two templates. This requires us to uniquely determine a type for every template that we generate. The test for template type equivalence is graph isomorphism.

The general graph isomorphism problem has never been proven NP-complete; however all proposed solutions to the problem possess exponential running time. The isomorphism problem itself is in the class of Gamma-NP problems, which have not yet been classified as being in either P or NP. Polynomial algorithms for certain classes of graphs, such as trees or planar graphs, are known. Unfortunately, little is known about isomorphism of directed acyclic graphs.

To perform type comparisons for templates, we integrated a polynomial-time approximation of DAG isomorphism into our implementation, via the University of Naples' VFLib Graph Matching Library [267]. The VFLib offers a selection of five different isomorphism algorithms: Ullman [268], Nauty [269], Schmidt and Druffel (DF) [270], VF [271], and VF2 [272]. A performance comparison of these algorithms is described in [267].

We chose the VF2 algorithm, based on the conclusion of [267]. In the performance evaluation in, SD, VF, and VF2 were the only algorithms to terminate. VF2 is similar to VF, but with reduced space complexity; furthermore, VF2 exhibited the best performance on smaller and sparser graphs, which exhibit similar characteristics to our DFGs. Although VF2 is an imperfect approximation (i.e. it will miss some DAGs that are isomorphic to one another) we found it to perform acceptably for our purposes. The worst-case time complexity of VF/VF2 was reported to be $O(V!V)$ [271]. VF2 is a re-implementation of VF with reduced space complexity from $O(N^3)$ to $O(N)$ [272].

4.4   Overlapping Templates

Both the Sequential and Parallel Template Generation Algorithms generate a set of candidate vertex-edge-vertex triplets (Sequential) or vertex-vertex pairs (Parallel). If a vertex v is contained in more than one pair or triplet, we are faced with a complicated decision involving template clustering. After we cluster the first pair or triplet containing v, v is no longer a vertex in the graph, and the additional pairs containing v cannot be clustered. Selecting which pairs or triplets should be clustered and which shouldn't in such a way that the number of templates clustered is maximized is an NP-complete problem. The problem of selecting a template from a set of overlapping templates is illustrated in Figure 38.

In [262], overlapping templates are modeled as a conflict graph and the set of templates selected for clustering is the maximum independent set (MIS) of the conflict graph. The authors of [260] present a more complicated problem formulation as an integer-linear program that chooses templates such that the estimated gain of all templates chosen is maximized. Estimating the possible gain of a template at the compiler level is unrealistic; therefore, a compiler implementation of this problem must rely on an interface with lower-level tools.

We present an alternative approach: to cluster all vertices adjacent to v into a large template with v. This works quite well for the Sequential Algorithm, because it allows for larger templates to be clustered in a single operation (see Figure 39). The drawback in applying this approach to the Sequential Algorithm is that larger templates may occur with less frequency than smaller ones.

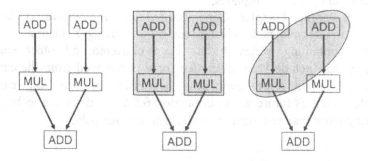

Figure 38 (a) a DFG; (b) the DFG from (a) with an optimal covering of (ADD, MUL) templates; (c) the same DFG with a sub-optimal covering due to poor selection among overlapping templates.

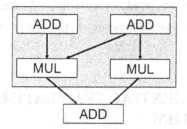

Figure 39 The example DFG from Figure 38 where all overlapping templates have been merged with one another.

This approach, unfortunately, cannot be used with the Parallel Template Generation Algorithm. Consider, for example, three vertices X, Y, and Z in Figure 40. X and Y can be scheduled at the same time step, and so can X and Z; however Y and Z cannot be scheduled at the same time step. Two possible templates are X-Y, and X-Z, and obviously both templates overlap at y; however, there is no legal clustering of template X-Y-Z that satisfies scheduling constraints. Therefore, overlapping templates cannot be merged when consider parallel templates.

For the Parallel Template Generation Algorithm, we have no choice but to select the templates to cluster one-by-one using a heuristic. Once a template containing vertex v has been clustered, all other candidate templates that include v are discarded (or v is removed from the candidate template, if applicable). Evaluating heuristics for clustering candidate templates is left as future work. In Section 6.5.5, we discuss a technique for clustering more than two parallel vertices simultaneously.

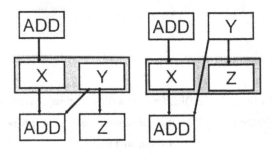

Figure 40 Y and X-Z may be scheduled in parallel; Y-Z cannot. Therefore merging overlapping parallel templates is impossible in certain cases.

5.   THE SEQUENTIAL TEMPLATE GENERATION ALGORITHM

5.1   Description of the Algorithm

Template generation was first applied to reconfigurable systems in [262]. This algorithm was capable of generating sequential templates; parallel templates and parallel regularity were not addressed. The algorithm is presented for reference in Figure 41. The function profile_graph traverses the DFG, G, and creates a frequency distribution of the edges that it observes by classifying them by type. The return value of profile_graph is the edge types ordered by their corresponding frequency. The function cluster_common_edges clusters vertices of the most frequently occurring edge type(s). Overlapping templates are dealt with as described in Section 4.4. The function stop_conditions_met controls termination of the algorithm. The criterion we selected for stop_conditions_met is to continue clustering until the most frequently occurring edge type in the DFG occurs only once.

Algorithm: Sequential Template Generation
Inputs: A dataflow graph $G = (V, E)$
Variables: C – a set of edge types

```
while(!stop_conditions_met(G))
  C = profile_graph(G)
  cluster_common_edges(G, C)
```

Figure 41 The Sequential Template Generation Algorithm

5.2   Detailed Example

In this section, we present a detailed example of the Sequential Template Generation Algorithm. We perform the algorithm on a DFG taken from the source code to the DeCSS encryption algorithm. We show intermediate results of the algorithm following each iteration as well as the edge type frequency distribution. Once the algorithm is complete, we analyze the modified DFG and describe a pipelined implementation of the templates in hardware.

We compiled the body of the following loop taken from the DeCSS algorithm:

```
for(i = 9; i >= 0; i--)
  key[CSStab0[i + 1]]  =  k[CSStab0[i + 1]] ^
                          CSStab1[key[CSStab0[i + 1]]] ^
                          key[CSStab0[i]]];
```

The DFG and the edge type frequency distribution for the computation are shown on the left side of Figure 42. We refer to this DFG as $G_0 = (V_0, E_0)$. The most frequently occurring edge type in E_0 is Add-Lod, which occurs 7 times. None of these templates overlap one another, so all 7 are selected for clustering. The resulting graph, $G_1 = (V_1, E_1)$, is shown on the right side of Figure 42. Additionally, graphs G_2 and G_3 are shown in Figure 43 and G_4 is shown on the left side of . Each figure shows the graphs, as well as their respective edge type frequency distributions. Every edge type in G_4 occurs only once, so the algorithm terminates after the fourth iteration.

Overlapping templates emerges as a problem during the second iteration, from G_1 to G_2. A total of 4 (Add-Lod, Add-Lod) edges are observed; however, two of the potential templates overlap. We are faced with two alternatives. One the one hand, we can either merge the overlapping templates into one larger template of type $(Add-Lod)^3$, giving us two templates of type $(Add-Lod)^2$ and one of $(Add-Lod)^3$. Alternatively, we can select one of the two overlapping templates, yielding three templates of type $(Add-Lod)^2$.

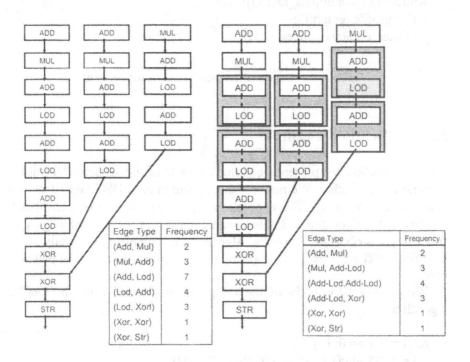

Figure 42 Original Graph G_0 (left) and Derived Graph G_1 (right)

We are faced with 3 possibilities: merge the templates, or select one or the other of the two overlapping templates. If we chose either of the other two possibilities, the edge type frequency distribution in G_2 would be radically different, and the algorithm would proceed accordingly. The final clustered graph could have a radically different template covering, and the algorithm may or may not terminate after the same number of iterations.

In the future, it may be possible to explore alternative clustering schemes solutions by utilizing a design space approach. Doing so would cause the time and/or space complexity of the algorithm to become exponential due to the size of the search space. Possible pruning techniques could be used to reduce these complexities; however, a tradeoff exists between efficiency in search and the quality of a solution. Additionally, optimization criteria must be designed to guide the search. Modeling the problem as a search space is beyond the scope of this chapter.

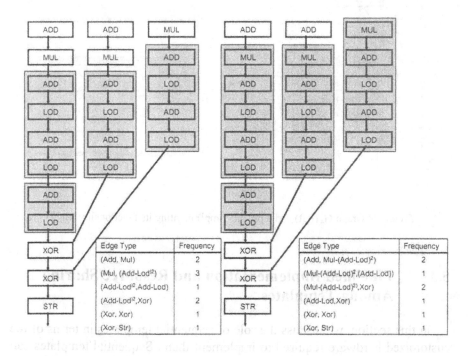

Figure 43 Graph G_2 (left) and Graph G_3 (right) along with edge type frequencies

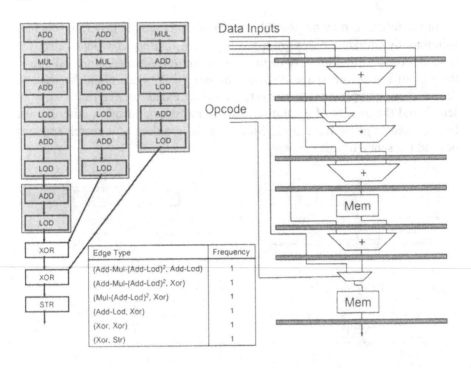

The following table appears within the figure:

Edge Type	Frequency
(Add-Mul-(Add-Lod)², Add-Lod)	1
(Add-Mul-(Add-Lod)², Xor)	1
(Mul-(Add-Lod)², Xor)	1
(Add-Lod, Xor)	1
(Xor, Xor)	1
(Xor, Str)	1

Figure 44 Graph G_4 (left) and a pipeline implementing its functionality (right)

5.3   Pipelined Implementation and Resource Sharing Among Templates

In this section, we discuss the role of sequential templates in terms of the customized hardware required to implement them. Sequential templates can be implemented as a pipeline, assuming a storage element (pipeline register) is placed at every internal edge in the template. Using this structure, we observe that by increasing the complexity of the pipeline and augmenting specific stages with multiplexers and selection lines, we can allow a single pipeline to implement the functionality of multiple templates.

A pipeline capable of implementing all template functionality for the clustered DFG shown on the left side of Figure 44 is shown on the right side of the figure. This pipeline is capable of implementing all three template operations: Add-Lod, Mul-Add-Lod-Add-Lod, and Add-Mul-Add-Lod-Add-Lod. If we assume that the initial pipeline was for the Add-Mul-Add-Lod-Add-Lod operation, then the two other instructions can be implemented using the pipeline at a cost of two multiplexers and two op-code inputs. The number of cycles required to compute each of the three instructions on the pipeline is equal to the number of internal instructions. That is, Add-Lod operates in 2 cycles, Mul-Add-Lod-Add-Lod operates in 5 cycles, and Add-Mul-Add-Lod-Add-Lod operates in 6 cycles. This is equivalent to the latency we would expect if we implemented 3 pipelines, one for each instruction, without the considerable area and storage overhead.

5.4 Complexity Analysis of the Sequential Template Algorithm

In [262] it was shown that one iteration of the Sequential Template Generation Algorithm requires $O(|E|)$ time. In this section, we present a more rigorous and more formalized analysis of the time complexity of the Sequential Template Generation Algorithm. In the analysis in [262], an important detail was overlooked. During each iteration of the algorithm, each edge in the DFG is examined to create the type frequency distribution. Either, some subset of the edges will be removed from the DFG if clustering occurs, or the algorithm terminates. If the algorithm does not terminate, the next iteration will proceed analogously to the first, however, the number of edges in the DFG will be reduced. The number of edges counted during each iteration is certainly bounded from above by $|E|$. We believe that a tighter analysis can bound the number of edges counted in the ith iteration by the number of edges remaining in the DFG after i-1 iterations.

Any realistic complexity analysis must take into account the heuristics used for clustering approaches and stopping conditions. We present an analysis for an implementation of the Sequential Template Generation described in Section 5.1. The algorithm terminates when the most frequently occurring edge type occurs once or when a single template subsumes the entire DFG. We assume that the problem of overlapping templates is resolved in constant time, and that clustering a template takes constant time as well. We do not consider the time complexity of isomorphism in this analysis.

We begin with the following definitions. Let $G = (V, E)$ be a DFG. Let $G_i = (V_i, E_i)$ be G after i iterations of the Sequential Template Generation Algorithm. We assume that the algorithm begins with a DFG $G_0 = (V_0, E_0)$ defined to be G. For now, we assume that the algorithm terminates after n iterations. We will quantify n later in this section.

First, we derive an upper bound on the number of iterations of the Sequential Template Generation Algorithm.

Lemma 7.1: $|V_i| < |V_{i-1}|$ for i = 1, 2, ..., n

Proof: Assume to the contrary that $|V_i| \geq |V_{i-1}|$. If $|V_i| > |V_{i-1}|$, then the ith iteration of the Sequential Template Generation Algorithm increased the number of vertices in the DFG. This cannot happen because the Sequential Template Generation Algorithm never performs a reverse-clustering operation, which cannot occur. Additionally, the algorithm introduces no external vertices at any point. This leads to a contradiction.

Now, if $|V_i| = |V_{i-1}|$, then no vertices were clustered during the ith iteration. Therefore, the most frequently occurring edge type in E_{i-1} occurs just once or $|E_{i-1}| = 0$. In either case, then the algorithm would terminate following the (i-1)st iteration. This contradicts the assumption that an ith iteration occurred. ❏

Lemma 7.2: $|V_i| \leq |V_{i-1}| - 2$, for $i = 1, 2, \ldots, n$.

Proof: $|V_i| < |V_{i-1}|$ by Lemma 1. Therefore, $|Vi| \leq |V_{i-1}| - 1$ since the number of vertices in the DFG is a discrete quantity. Now, assume that $|Vi| = |V_{i-1}| - 1$. This indicates that during the ith iteration of the Sequential Template Generation Algorithm, two vertices v_1 and $v_2 \in V_{i-1}$ were removed and replaced with a template t. Therefore:

$$V_i = V_{i-1} - \{v_1, v_2\} \cup \{t\}$$
$$|V_i| = |V_{i-1}| - 1$$

Since only v_1 and v_2 are clustered, then the most frequently occurring edge type only occurred once in E_{i-1}. If this occurred, then the algorithm would have terminated after the (i-1)st iteration, which is a contradiction. Hence, $|Vi| < |Vi-1| - 1$, which we express as $|Vi| \leq |Vi-1| - 2$. ❏

Lemma 7.3: $|E_i| \leq |E_{i-1}| - 2$, for $i = 1, 2, \ldots, n$.

Proof: Each time a pair of vertices is replaced by a template, exactly one edge is subsumed since templates are identified based on edge type. Given this, the result follows directly from Lemma 7.2. ❏

Lemma 7.4: The Sequential Template Generation Algorithm terminates after $n \leq |V|/2$ iterations.

Proof: We consider the situation wherein every vertex in V is eventually subsumed by a template. If $|V|$ is odd, then the algorithm terminates with $G_n = (\{t_1, t_2\}, \{(t_1, t_2)\})$, where t_1 and t_2 are templates. If $|V|$ is even, then the algorithm terminates with Gn = $(\{t\}, \phi)$, where t is a template. n is maximized when the number of vertices replaced by template(s) during each iteration is minimized. By applying Lemma 7.2, we see that the number of vertices in the DFG is reduced by at least 2 every iteration. When this occurs, the condition described above will occur after exactly $|V|/2$

iterations. Thus, n = |V|/2 in the worst case. In general, we conclude that n ≤ |V|/2. ❑

Lemma 7.5: The ith iteration of the Sequential Template Generation Algorithm has time complexity $O(|V_i| + |E_i|) = O(|V| + |E|)$.

Proof: From Lemma 7.2, we can trivially establish that $|V_i| \leq |V| - 2i$. In the worst-case scenario described in the proof of Lemma 4, $|V_i| = |V| - 2i$. From Lemma 3, we can trivially establish that $|E_i| \leq |E| - 2i$. In the worst-case scenario, $|E_i| = |E| - 2i$.

The stop condition and profiling function are analogous to one another: both rely on a frequency distribution of edge types, which can be calculated in $O(|E_i|)$ time. By Theorem 1, Clustering requires $O(|V_i| + |E_i|)$ time as well. The total time complexity of one iteration of the Sequential Template Generation Algorithm is thus $O(|V_i| + |E_i|)$. Since $|V_i| = |V| - 2i$ and $|E_i| = |E| - 2i$ this becomes $O(|V| + |E|)$ in the worst-case. ❑

Theorem 7.3: The total time complexity of the Sequential Template Generation Algorithm is given by $T(V, E) = O(|V\|E|)$.

Proof: We combine the results of Lemma 7.4 and Lemma 7.5. $T(V, E)$ is computed as follows:

$$T(V, E) = \sum_{i=0}^{n}(|V_i| + |E_i|) = \sum_{i=0}^{n}(|V| + |E| - 4i) = n(|V| + |E|) - 4\sum_{i=0}^{n}i$$

$$= n(|V| + |E|) - 2n(n+1) \leq \frac{|V|}{2}(|V| + |E|) - 2\frac{|V|}{2}\left(\frac{|V|}{2} + 1\right)$$

$$= \frac{|V|}{2}[(|V| + |E|) - (|V| + 2)] = \frac{|V|}{2}(|E| - 2) = O(|V\|E|) \ ❑$$

The above result does not contradict the analysis of [262]. The contribution of this section is to describe the precise behavior of the algorithm and provides a tighter bound on the per-iteration time complexity. Additionally, we derived a total time complexity for the entire algorithm, which was not addressed in [262].

6.   THE PARALLEL TEMPLATE GENERATION ALGORITHM

6.1   Overview

We describe the Parallel Template Generation Algorithm [263] as an alternative to the Sequential Template Generation Algorithm. The Parallel Algorithm is similar in structure to the Sequential Algorithm; however, the approach to profiling is different. Instead of counting DFG edges like its Sequential counterpart, the Parallel Algorithm first constructs an interval graph called the All-Pairs Common Slack Graph (APCSG). An edge exists between two vertices in the APCSG if and only if the two vertices could be scheduled in parallel (at the same time-step) by some latency-constrained scheduling heuristic acting upon the DFG. The Parallel Algorithm profiles the APCSG instead of the DFG. APCSG edges, unlike DFG edges, are undirected. The only difference in the profiling involves computing types for directed and undirected edges, which are described in Section 4.3.

Templates generated by the Parallel Algorithm have different structure than the templates generated by the Sequential Algorithm. Parallel templates are data independent, unlike Sequential Templates. As a result, it is possible to schedule the operations subsumed by the template at the same time step, assuming the availability of the required functional resources. Hence, Parallel Template Generation may be used to extract operation level parallelism from a DFG.

6.2   Construction of the APCSG

In this section, we describe how to construct the APCSG from a DFG. The algorithm is presented in FIGURE 16. The crux of the algorithm is this: in the APCSG, an edge is placed between two vertices if and only if the corresponding vertices in the DFG could possibly be scheduled at the same time step by some scheduling heuristic. For this to occur, two properties must hold: there may be no data dependence between the vertices, and the common slack between the vertices must be positive.

We formalize our notion of common slack as follows. Let $G = (V, E)$ be a DFG, and let v be a vertex of G. We define the level of v, $L_\chi(v)$ to be the time step at which v is scheduled by some latency constrained scheduling heuristic χ. For our purposes, the only scheduling heuristics we are concerned with are the As-Soon-As-Possible (ASAP) and the As-Late-As-Possible (ALAP) Heuristics.

The slack of a DFG vertex is defined to be the difference between the levels at which it is scheduled by the ALAP and ASAP heuristics. That is, for any vertex v: $Slack(v) = L_{ALAP}(v) - L_{ASAP}(v)$. $L_{ALAP}(v)$ is the latest possible time step at which v can be scheduled; $L_{ASAP}(v)$, in contrast, is the earliest possible time step. Slack is a nonnegative quantity. Any vertex v such that $Slack(v) = 0$ is said to be on the critical path of the graph. In this case, $L_\chi(v)$ is fixed for all latency-constrained scheduling heuristics.

We define the common slack of a pair of vertices u and v, $CS(u, v) = CS(v, u)$ to be the number of levels at which u and v may be scheduled together by some latency-constrained scheduling heuristic. All vertices that may not be scheduled together have common slack values of zero.

Computing common slack entails a two-part process. First, we observe that no two vertices share common slack if they are data dependent. Data dependence information for any pair of vertices in a DFG G can be obtained by computing the transitive closure of G, denoted TC_G.

The transitive closure is a binary $|V| \times |V|$ matrix. For any pair of DFG vertices u and v, $TC_G(u, v) = 1$ if and only if there is a path from u and v in G; 0 otherwise. $TC_G(u, u) = 1$ by definition, for all vertices u. If $TC_G(u, v) = 1$, then $CS(u, v) = 0$. If $TC_G(u, v) = 0$, however, it is possible, but not guaranteed, that $CS(u, v) > 0$; that is, that u and v have common slack.

Algorithm: APCSG Construction
Inputs: A dataflow graph G = (V, E), Transitive Closure of G, TC_G
Output: A weighted interval graph $G_{APSCG}(V, E_{APCSG}, w)$
Variables: Matrix of common slack values, CS

$E_{APCSG} = \{\}$
Compute $L_{ASAP}(v)$ and $L_{ALAP}(v)$ for every vertex v using topological sort.
For every vertex $v \in V$
    $Slack(v) = L_{ALAP}(v) - L_{ASAP}(v)$
For every pair of vertices u, $v \in V$
Select u, v, such that $L_{ASAP}(u) \leq L_{ASAP}(v)$
If $TC_G(u, v) = 0$ and $L_{ALAP}(u) \geq LASAP(v)$
    $CS(u, v) = min(L_{ALAP}(u) - L_{ALAP}(v) + 1, Slack(v) + 1)$
    Add an edge e = (u, v) to E_{APCSG}
    Set weight function w(u, v) = CS(u ,v)
    Else
  $CS(u, v) = 0$
Return GAPSCG = (V, EAPCSG, w)

Figure 45 The APCSG Construction Algorithm

Without loss of generality assume that $L_{ASAP}(u) < L_{ASAP}(v)$. If $L_{ALAP}(u) < L_{ASAP}(v)$, then u and v cannot be scheduled at the time step by any heuristic despite the fact that they are data independent. Thus, $CS(u, v) = 0$. On the other hand, if $L_{ALAP}(u) \geq L_{ASAP}(v)$, then $CS(u, v) > 0$..

If our goal was simply to construct an interval graph, we would add an edge $e = (u, v)$ to the APCSG at this point and move on to the next pair of vertices; however, we are interested in the amount of slack shared between two vertices. That is, we desire the edges of our APCSG to be weighted with the common slack between the two vertices incident on the edge. Assuming that $TC(u, v) = 0$ and $L_{ALAP}(u) \geq L_{ALAP}(v)$, $CS(u, v)$ is computed as follows:

$CS(u, v) = \min(L_{ALAP}(u) - L_{ALAP}(v) + 1, Slack(v) + 1)$

For every pair of vertices u and v in a DFG G, we add to the APCSG a weighted edge (u, v) with weight function $w(u, v) = CS(u, v)$ if and only if $CS(u, v) > 0$. Equivalently, we may use CS as an adjacency matrix representation of the APCSG.

6.3   The Parallel Template Generation Algorithm

The Parallel Template Algorithm is presented in Figure 46. The algorithm takes one parameter, a DFG $G = (V, E)$. First, the APCSG Construction Algorithm described in the previous section is applied to G, yielding G_{APCSG}. Next, the algorithm iterates until a sufficient stopping condition is met.

Each iteration begins by profiling the G_{APCSG}, which entails the task of creating a frequency distribution, C, of edge types in E_{APCSG}. In the next step, we create parallel templates in G. Any pair of vertices u and v connected by an edge (u, v) or (v, u) in E_{APCSG} may be clustered as parallel templates. In order to maximize regularity, we consider clustering u and v together if and only if $t((u, v))$ is the most frequently occurring edge type in C. Because of dependency issues, we cannot cluster every such pair of vertices. Selection heuristics are discussed in Section 6.5.

Finally, we must update G_{APCSG}. Clustering a pair of vertices in parallel may alter the transitive closure of G, introducing new dependencies between previously independent vertices in V. Additionally, a vertex that is a parallel template may have a slack value smaller than either of the vertices it subsumed in a previous iteration. As a result, the APCSG must be updated or recomputed. Once the APCSG is updated, the stopping conditions are evaluated once again, and the algorithm iterates or terminates accordingly.

Algorithm: Parallel Template Generation
Inputs: A DFG G = (V, E)
Variables: An APCSG $G_{APCSG}(V, E_{APCSG})$
      Transitive Closure of G, TC_G
      C – a set of APCSG edge types

TC_G = Compute_Transitive_Closure(G)
G_{APCSG} = construct_apcsg(G, TC_G)
while(stop_conditions_met(G, G_{APCSG}))
  C = profile_graph(G_{APCSG})
  cluster_parallel_templates(G, C)
  update_apcsg(G, TC_G, G_{APCSG})

Figure 46 The Parallel Template Generation Algorithm

6.4 Updating the APCSG

In this section, we discuss the problem of how to update the APCSG during each iteration of the algorithm. One solution would be to discard the APCSG after each iteration, and create a new APCSG for the next; however, APCSG construction entails computing transitive closure, which is itself an $O(V^3)$ algorithm, which is too costly. Instead, we propose a mechanism to update the APCSG that does not require re-computing the transitive closure.

We desire a mechanism to create $T_{G'}$ from T_G that is more efficient than the $O(V^3)$ transitive closure algorithm. Additionally, we must consider situations in which we cluster more than one template at the same time.

Given these observations, we propose a technique to update the APCSG after each iteration of the Parallel Template Generation Algorithm. We begin with several definitions. Let G = (V, E) be a DFG. Let $G_i = (V_i, E_i)$ be G after i iterations of the Parallel Template Generation Algorithm. We assume that the algorithm begins with a DFG $G_0 = (V_0, E_0)$ defined to be G. We define the APCSG and transitive closure similarly. Let $G_{APSCG} = (V, E_{APCSG}, w)$ be the APCSG of G. Let $G_{APCSG,i} = (V_i, E_{APCSG,i}, w_i)$ be the APCSG after i iterations of the Parallel Template Generation Algorithm. We define $G_{APCSG,0} = (V_0, E_0, w_0)$ to be G_{APCSG}. Finally, let TC_G be the transitive closure of G. Let $TC_{G,i}$ be the transitive closure of G_i.

We observe that if there is no edge between vertices u and v in $G_{APCSG,i}$, then no edge will be placed betweens vertices u and v in $G_{APCSG,i+k}$, $k \geq 0$. Therefore, when creating $G_{APCSG,i+1}$, we only need to adding edges between vertices that were adjacent in $G_{APCSG,i}$. Unfortunately, we cannot tell if two vertices that are adjacent in $G_{APCSG,i}$ are data dependent in G_{i+1} without updating the transitive closure.

First, we solve the problem of updating the transitive closure in less than $O(V_3)$ time. Second, we show how to update the APCSG by utilizing the updated transitive closure.

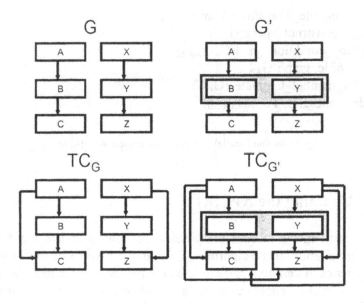

Figure 47 A DFG G and its transitive closure TC_G, and a DFG G' which is G with a parallel

template, and the resulting transitive closure, $TC_{G'}$

6.4.1   Updating Transitive Closure

The problem of updating the transitive closure is defined as follows: Given $TC_{G,i}$ and G_{i+1}, compute $TC_{G,i+1}$. To simplify our discussion, we consider the simplest case: when we replace exactly two vertices with one parallel template. This algorithm is presented in Figure 48. This algorithm could easily be extended to support the clustering of more than one template simultaneously, or templates consisting of more than 2 vertices.

The update procedure is described as follows. For illustrative purposes, we replace vertices u and v with vertex t in G_i. This yields G_{i+1}. Second, we note that $TC_{G,i+1}$ will contain an edge (x, t) if and only if $TC_{G,i}$ contains an edge (x, u) or (x, v); similarly, $TC_{G,i+1}$ will contain an edge (t, y) if and only if $TC_{G,i}$ contains an edge (u, y) or (v, y). This computation is achieved in lines 3 – 5 of the algorithm. Next, we consider every pair of vertices x and y other than t in $TC_{G,i+1}$. If a (x, y) or a (y, x) edge exists in $TC_{G,i}$, then the edge is propagated to $TC_{G,i+1}$; regardless of whether or not the x-y/y-x paths that occur in $TC_{G,i}$ utilize u or v; if no such edge exists, we must determine if such an edge should be added as a result of clustering u and v together. If no x-y edge exists in $TC_{G,i}$, then an x-y edge is added to $TC_{G,i+1}$ if the following two conditions hold:

(1) $\exists$ (x, u) or (x, v) $\in TC_{G,i}$ $\Leftrightarrow$ $\exists$ (x, t) $\in TC_{G,i+1}$

(2) $\exists$ (u, y) or (v, y) $\in TC_{G,i}$ $\Leftrightarrow$ $\exists$ (t, y) $\in TC_{G,i+1}$

If both of these conditions hold, then an edge (x, y) is added to $TC_{G,i+1}$. After considering every pair of vertices, the update terminates.

Algorithm: Update Transitive Closure
Inputs: Old Transitive Closure, TCG,i
    DFG $G_i = (V_i, E_i)$
    u, v $\in$ Vi – pair of vertices to be clustered.
Variables: t – the template replacing vertices u and v.
Output: New Transitive Closure $TC_{G,i+1}$

$V_{i+1} = V_i - \{u, v\} \cup \{t\}$
$E_{i+1} = E_i - \{(x, y) \mid (x = u \text{ or } v) \text{ or } (y = u \text{ or } v)\}$
        $\cup \{(z, t) \mid (z, u) \text{ or } (z, v) \in Ei \}$
        $\cup \{(t, z) \mid (u, z) \text{ or } (v, z) \in E_i \}$
For each vertex x $\in$ Vi, x $\neq$ u, v
    $TC_{G,i+1}(x, t) = TC_{G,i}(x, u) \mid TC_{G,i}(x, v)$
    $TC_{G,i+1}(t, x) = TC_{G,i}(u, x) \mid TC_{G,i}(v, x)$
For each pair of vertices x, y $\in$ Vi, x, y $\neq$ u, v
    $TC_{G,i+1}(x, y) = TC_{G,i}(x, y) \mid (TC_{G,i+1}(x, t) \& TC_{G,i+1}(t, y))$
    $TC_{G,i+1}(y, x) = TC_{G,i}(y, x) \mid (TC_{G,i+1}(y, t) \& TC_{G,i+1}(t, x))$

Figure 48 Algorithm for Updating the Transitive Closure

6.4.2   Updating the APCSG

Once we have successfully updated the transitive closure, we must then update the APCSG. The algorithm is presented in FIGURE 30. This algorithm takes several parameters: three graphs: G_{i+1}, $TC_{G,i+1}$, and $G_{APCSG,i}$, and two vertices, u and v, and a parallel template t, which will replace them.

First, we recompute the slack values for each pair of vertices using topological sort. Second, we consider every edge (x, y) in $E_{APCSG,i}$. If clustering u and v has introduced a dependency between x and y, then no edge (x, y) can be added to $E_{APCSG,i+1}$; otherwise, we compute the common slack between x and y, and add (x, y) to $E_{APCSG,i+1}$ if and only if x and y have positive common slack in G_{i+1}. This second step is quite similar to the basic step used for APCSG construction. Once this step is completed for every pair of vertices, the update algorithm terminates and returns $G_{APCSG,i+1}$.

Algorithm: Update APCSG
Inputs: New Transitive Closure, TCG,i+1
DFG $G_{i+1} = (V_{i+1}, E_{i+1})$
Old APCSG, $G_{APCSG,i} = (V_i, E_{APCSG,i}, w_i)$
u, v $\in$ Vi – pair of vertices to be clustered
t – template vertex replacing u and v.
Output: New APCSG $G_{APCSG,i+1} = (V_{i+1}, E_{APCSG,I+1}, w_{i+1})$

$V_{i+1} = V_i - \{u, v\} \cup \{t\}$
$E_{APCSG,i+1} = w_{i+1} = \phi$
Compute $L_{ASAP}(x)$ and $L_{ALAP}(x)$ for every vertex $x \in V_{i+1}$ by applying
    topological sort to G_{i+1}.
For each edge $(x, y) \in E_{APCSG,i}$
Select x, y, such that $L_{ASAP}(x) \leq L_{ASAP}(y)$
If $TC_{G,i+1}(x, y) = 0$ and $L_{ALAP}(x) \geq LASAP(y)$
$CS(x, y) = min(L_{ALAP}(x) - L_{ALAP}(y) + 1, Slack(y) + 1)$
$EAPCSG,i+1 = EAPCSG,i+1 \cup \{(x, y)\}$
$w_{i+1}(u, v) = CS(x, y)$
Return $G_{APCSG,i+1} = (V_{i+1}, E_{APCSG,I+1}, w_{i+1})$

Figure 49 Algorithm for Updating the APCSG at the end of each iteration of the Parallel
Template Generation Algorithm

6.5   Template Selection

In this section, we address the problem of selecting a subset of templates from a set of candidate templates such that no vertex appears more than once in the selected subset. We describe several different approaches for selection and contrast them with the approaches we used for the Sequential Template Generation Algorithm. Due to scheduling constraints, we cannot simply merge overlapping templates without (sometimes drastically) increasing the latency of the DFG. Therefore, this technique is not considered.

6.5.1   Problem Statement

We define the problem of selection among overlapping templates as follows. Let $T = \{T_1, T_2, ..., T_k\}$ be a set of candidate templates. Let $T_i = (x_i,$

y_i) be a candidate template that would join vertices x_i and y_i as a parallel template. We have the following restrictions on the vertex types described as follows:

$t(x_i) = t(x_j) \ \forall \ i, j, \ 1 \le i, j \le k$

$t(y_i) = t(y_j) \ \forall \ i, j, \ 1 \le i, j \le k$

$\exists \ x \in \{x_1, x_2, ..., x_k\}, y \in \{y_1, y_2, ..., y_k\}, \ni t(x) = t(y) \Leftrightarrow$
$t(x_i) = t(y_i) \ \forall \ i, j, \ 1 \le i, j \le k$

Criterion (1) and (2) ensure that all of the x's and y's respectively have the same types. Criterion (3) covers the case where both the x's and the y's share the same type (e.g. considering Add-Add templates). The goal is to select a maximal subset of the T_i's such that no vertex appears more than once among any of the selected T_i.

6.5.2   Non-Applicability of Set Cover

Based on the problem statement, one may mistakenly assume that template selection is a special case of the Set Cover Problem, which is known to be NP-Complete; however, template selection has several nuances that prevent either of these approaches from being applicable in the general case. The first of these two nuances derives from the fact that clustering a pair of vertices in parallel can create new dependencies between candidate template pairs. Additionally, clustering a pair of vertices in parallel can restrict the slack of other vertices. Consequently, common slack between certain vertex pairs can be eliminated. As a result, solutions to Template Selection based on Set Cover may not be feasible.

6.5.3   A Greedy Solution

As a consequence of the previous discussion, we hypothesize that the best solution is to select one pair of vertices at a time to cluster as a parallel template, and then update the transitive closure and APCSG before selecting another pair. Such an approach is greedy, because only one vertex at a time may be selected and no backtracking occurs. Obviously, a search-based approach could find an optimal solution based on any objective criterion one desired; we leave the exploration of search-based solutions to this problem as future work. Here, we address the problem of selecting which pair of vertices to cluster first from a set of candidates (some of which can be assumed to be eliminated as a result). We assume that candidate pairs are stored in an easily traversable list, and that the greedy approach is used.

The simplest approach is to randomly select a pair from the list of candidates. The easiest implementation of the randomized approach is to always select the first candidate in the list, assuming that the list has random order. If this approach is used, selection can be accomplished in O(1) time at each step; however, no assumptions can be made about the quality of a solution.

A more sophisticated approach is to compute an objective value $Obj(T_i)$ for each template T_i, and select template T_i such that $Obj(T_i)$ is maximized or minimized accordingly. Assuming that $Obj(T_i)$ can be computed in $O(1)$ time, this approach requires $O(|T|)$ time to examine all candidates. The effectiveness of this approach is heavily dependent on the choice of objective function. Furthermore, if the objective function requires considerable time to compute, the greedy approach can become unwieldy and inefficient; a search-based backtracking approach may offer better results at an equivalent or slightly more expensive cost.

6.5.4   Our Objective Function

In the field of optimization theory, academics may argue indefinitely over the use of objective functions. Often, there is no right answer; other times, the answer changes over time as new objective functions are proposed, tested, and refined. Often, the best objective function at any given time is the one that has yielded the best results in the most recent literature. Here, we describe the objective function that we selected and attempt to justify our choice. It would come as no surprise to us if better objective functions are discovered in future research experiments (either by us or by others).

Our choice of objective function was driven by our desire to integrate template generation into a larger high-level synthesis framework. In this framework, template generation is performed prior to (or alternatively, as the first step in) high-level synthesis. Following template generation are the traditional tasks of allocation, scheduling, resource binding, and resource selection. We observed that clustering a pair of vertices as a parallel template is in many ways analogous to scheduling their respective operations at the same time step. Further reasoning led us to believe that a poor choice of template selection could restrict the quality of schedule achieved during high-level synthesis. As a result, our goal in selecting an objective function was to identify one that would minimize the negative effects of scheduling.

The goal of our objective function, in other words, is to minimize the slack reduction realized from clustering a pair of templates. Let x and y be a pair of templates being clustered, with slack values Slack(x) and Slack(y), and common slack CS(x, y) respectively. We define the slack reductions of vertex x (y) when clustered with vertex y (x) as follows:

Reduction(x) = Slack(x) – CS(x, y), and
Reduction(y) = Slack(y) – CS(x, y)

From a conceptual standpoint, our objective is to select a pair of vertices x and y to cluster as a parallel template such that minimizes a function combining Reduction(x) and Reduction(y). For example, we may wish to minimize any of the following:

$F_1(x, y) = \text{Reduction}(x) + \text{Reduction}(y)$

$F_2(x, y) = \text{Reduction}(x)^2 + \text{Reduction}(y)^2$

$F_3(x, y) = \text{Reduction}(x) \times \text{Reduction}(y)$

$F_4(x, y) = \text{Reduction}(x)^2 \times \text{Reduction}(y)^2$

As previously mentioned, the choice of ideal objective function is usually determined experimentally; over time, as more objective functions are considered, better and better ones typically emerge.

6.5.5   Clustering More than Two Vertices into a Single Template

One interesting observation is that we could generate parallel templates consisting of more than two vertices at a single time. The criterion for clustering a set of vertices, $\{v_1, v_2, ..., v_j\}$ together as a parallel template is that the vertices must form a clique in the APCSG; that is, there must be an APCSG edge between every pair of vertices in the set. In order to find regularly occurring templates satisfying this criteria, we would have to search for a maximal set of independent cliques covering an equivalent set of vertices (i.e. a multiset of vertex types). Given that this is a composition of two NP-Complete problems, we do not propose this as a realistic solution; however, if the size of a DFG is known to be small, such a technique could certainly be applicable. On the other hand, small DFGs exhibit minimal regularity, which make them poor candidates for regularity extraction.

6.5.6   Refining the Parallel Template Generation Algorithm

After the previous discussion in this section, we reformulate the parallel template generation algorithm in lieu of our template selection heuristic. In this advanced formulation, the APCSG (and transitive closure) are updated after every pair of vertices is clustered. The algorithm is shown in Figure 50, omitting the details of template selection—the basic assumption was that the set of templates to be clustered could be constructed, the templates could then be clustered, and the APCSG updated after the fact. The formulation presented here shows the greedy heuristic in action, and updates the APCSG after every pair of templates is clustered. Additionally, vertex pairs that are no longer legitimate candidates are removed from the list of candidates.

Algorithm: Parallel Template Generation
Inputs: A dataflow graph G = (V, E)
Variables: An APCSG $G_{APCSG}(V, E_{APCSG})$, TC_G - Transitive Closure of G
   and C – a set of APCSG edge types

1. TC_G = Compute_Transitive_Closure(G)
2. G_{APCSG} = construct_apcsg(G, TC_G)
3. while(stop_conditions_met(G, G_{APCSG}))
4.   C = profile_graph(G_{APCSG})
5.   while C is not empty
6.   select the most objective-efficient pair of vertices x and y
7.   update_apcsg(G, x, y, TC_G, G_{APCSG})
8.   remove all vertex pairs from C that contain x or y, or no longer share
         common slack; recalculate the common slack between all pairs of
         remaining vertices.

Figure 50 The Reformulated Parallel Template Generation Algorithm

6.6   Complexity Analysis

In this section, we present a complexity analysis of the Parallel Template Generation Algorithm. Many of the results of the complexity analysis of the Sequential Template Generation Algorithm apply to the Parallel Algorithm as well. We begin by analyzing the APCSG construction algorithm. We begin with a DFG G = (V, E) having respective transitive closure TC_G. It is a well-known result that the complexity of constructing TC_G is $O(|V|^3)$.

6.6.1   APCSG Construction Algorithm

Lemma 7.6: The Time Complexity of the APCSG Construction Algorithm is $O(|V|^2)$.

Proof: The complexity of performing topological sort is $O(|V|+|E|)$. The complexity for computing the slack value for all of the vertices in the DFG is $O(|V|)$ since slack can be computed in constant time. The complexity of computing the common slack between each pair of vertices is $O(|V|^2)$ since every combination of two distinct vertices must be examined and common slack can be computed in constant time. The total time complexity of constructing the APCSG is $O(|V|^2)$.

6.6.2   Profiling the APCSG

Lemma 7.7: The Time Complexity to profile the APCSG is $O(|E_{APCSG}|)$.

Proof: To create the frequency distribution of edge types, we must examine each edge in the APCSG exactly once. The complexity of this algorithm is trivially $O(|E_{APCSG}|)$. ❑

6.6.3   Updating the Transitive Closure

Refer to the algorithm in Figure 48. $G_i = (V_i, E_i)$ and $TC_{G,i}$ refer to the DFG and transitive closure before vertices u and v are clustered and replaced with template vertex t. $G_{i+1} = (V_{i+1}, E_{i+1})$ and $TC_{G,i+1}$ refer to the DFG and transitive closure afterward.

Lemma 7.8: The time complexity of updating the transitive closure for a single parallel template that subsumes two vertices is $O(|V_i|^2)$

Proof: Lines 1 and 2 of the algorithm are implicit: they describe how DFG G_{i+1} is created from G_i when a single two-vertex template is clustered. Vertices u and v must be removed from the list of vertices, and template vertex t must be added; similarly, every edge adjacent to u or v must be replaced with an edge adjacent to t instead. This clustering procedure can be accomplished in $O(|V_i| + |E_i|)$ time.

Next, we consider for-loops in lines 3-5 and 6-8 respectively. The first for-loop examines each vertex in V_i, other than u and v, and thus requires $O(|V_i|)$ time. The second for-loop involves examining pair of vertices in V_{i+1} other than the template vertex t. This requires $O(|V_{i+1}|^2)$ time. Since $|V_{i+1}| = |V_i| + 1 = O(|V_i|)$, the total time complexity is $O(|V_i|^2)$. ❑

Lemma 7.9: Given an updated DFG G_{i+1} and an updated transitive closure graph, $TC_{G,i+1}$, then the process of updating an old APCSG $G_{APCSG,i} = (V_i, E_{APCSG,i}, w_i)$ takes $O(|V_i| + |E_i| + |E_{APCSG,i}|)$ time.

Proof: Refer to the algorithm in FIGURE 30. Vi+1 can be computed from Vi in Line 1 in $O(|Vi|)$, as in Lemma 7.8. Recomputing the slack values in Line 2 by topologically sorting G_{i+1} takes $O((|Vi+1| + |E_{i+1}|)$ time. The for-loop in line 4 examines every edge in the old APCSG, requiring $O(|EAPCSG,i|)$ iterations. All of the operations inside of the for loop are basic $O(1)$ time operations. Therefore, the total time to update the APCSG, given an updated transitive closure, is $O(|Vi+1| + |E_{i+1}| + |EAPCSG,i|)$ time. $|Vi+1| = O(|Vi|)$ from the proof of Lemma 7.8, and similarly $|Ei+1| = O(|Ei|)$, the time complexity becomes $O(|Vi| + |Ei| + |EAPCSG,i|)$. ❑

Lemma 7.10: The combined process of updating the Transitive Closure and APCSG requires $O(|V_i|^2)$ time.

Proof: Immediate from Lemma 7.8 and Lemma 7.9. The $|Vi|2$ term from Lemma 7.8 dominates the $|Vi+1|$, $|E_{i+1}|$, and $|EAPCSG,i|$ terms in Lemma 7.9. ❑

6.6.4   The Process of Template Selection

The complexity of clustering a set of parallel templates depends on whether or not the templates overlap one another and the methodology used to select a subset of templates for clustering from a set of overlapping templates. In this analysis, we assume that we use solve this problem using the greedy heuristic described in Section 6.5.3 and that the object function can be computed for each candidate in constant time.

Lemma 7.11: Each iteration of the inner for-loop (lines 6-8) of the Parallel Template Generation Algorithm (FIGURE 32) requires $O(|V_i|^2)$ time.

Proof: The cost of determining the most objective-efficient pair of vertices in C in line 6 requires $O(|C|)$ time, since the objective value can be computed in $O(1)$ time for each candidate in the list. Similarly, removing the appropriate vertices from C in line 8 require $O(|C|)$ time. From Lemma 10, the cost of updating the APCSG in line 7 requires $O(|V_i|^2)$ time. Thus, the complexity of excuting lines 6-8 is $O(|C| + |V_i|^2)$. Furthermore, observe that $|C| \leq |V_i|^2$ since at most every pair of vertex is a possible candidate (in the case that the DFG contains no edges). Therefore, the time complexity of executing lines 6-8 becomes $O(|V_i|^2)$. ❑

6.6.5   Complexity of the Inner While-Loop

In this section, we concentrate on determining the complexity of the inner while-loop. Initially, we let $G_j = (V_j, E_j)$ be the DFG after the jth execution of line 4 of the outer while loop of the Parallel Template Generation Algorithm (FIGURE 32). We bound the number of iterations of the inner while-loop in Lemma 7.12. In Lemma 7.13, we show the complexity of the inner while-loop in terms of $|V_j|$.

Lemma 7.12: The inner while loop (line 5) iterates at most $|V_j|/2$ times.

Proof: After the inner while-loop (lines 5-8) terminates, a template vertex covers each DFG vertex in the worst case. Each template covers exactly two vertices, thus there is a total of $|V_j|/2$ templates. Since each iteration of the inner while loop (line 5) adds exactly one template to the DFG, the number of iterations is bounded from above by $|V_j|/2$. ❑

Lemma 7.13: The time complexity of the complete the inner while-loop is $O(|V_j|^3)$.

Proof: Each iteration of the inner while-loop removes 2 vertices from the graph and replaces them with a template vertex. Initially, $|C| = |V_j|^2$. After the first candidate is selected, $|C| = (|V_j| - 2)^2$. In general, after the kth iteration, $|C| = (|V_j| - 2k)^2$ The time complexity of the loop is given as follows:

$$T(V_j) = \sum_{i=0}^{\frac{|V_j|}{2}} (|V_j| - 2i)^2 \leq \sum_{i=0}^{\frac{|V_j|}{2}} |V_j|^2 = \frac{|V_j|}{2} \times |V_j|^2 = O(|V_j|^3) \; ❑$$

6.6.6 Complexity of the Outer-While Loop

In this section, we determine an upper bound on the number of iterations of the outer while loop and determine the time complexity for executing it.

Lemma 7.14: The outer while-loop iterates at most $\log|V|$ times.

Proof: In the worst case, the inner loop covers every vertex in the DFG with a template. Since each template covers exactly two vertices the number of vertices in the graph is halved. After $\log|V|$ iterations, only one vertex will be left in the DFG, and we can go no further. ❑

Lemma 7.15: The time complexity of the outer while-loop is $O(|V|^3\log|V|)$.

Proof: Given Lemma 7.14, we can compute the time complexity of the outer loop as follows:

$$T(V) = \sum_{i=0}^{\log V} \left(\frac{|V|}{2^i} \right)^3 = |V|^3 \sum_{i=0}^{\log V} \frac{1}{8^i} \leq |V|^3 |\log|V| = O(|V|^3 |\log|V|) \; ❑$$

6.6.7 Complexity of the Parallel Template Generation Algorithm

In this section, we prove our main result: the time complexity of the Parallel Template Generation Algorithm

Theorem 7.4: The Time Complexity of the Parallel Template Generation Algorithm is $O(|V|^3\log|V|)$.

Proof: Consider the Parallel Template Generation Algorithm in FIGURE 32. It is a well known result that computing the transitive closure of a DAG requires $O(|V|^3)$ time. From Lemma 6, the complexity of computing the APCSG is $O(|V|^2)$. From Lemma 15, the complexity of the outer while-loop (line 3) is $O(|V|^3\log|V|)$. Combining these results, we observe that the total time complexity is $O(|V|^2 + |V|^3 + |V|^3\log|V|) = O(|V|^3\log|V|)$. ❑

7.   EXPERIMENTS AND RESULTS

We implemented our algorithms on top of the Stanford SUIF compiler [273]. Building upon some of the modifications made by Kastner for sequential template construction [262], we added a full implementation of the APCSG and its generation, as well as the code to successfully merge parallel and sequential templates. Finally, we added the higher-level heuristic that performed the combination of sequential and parallel clustering. Our template sizes were restricted to five internal nodes, and our algorithm terminated when the total number of super-nodes (clustered vertices) in the DFG was less than half of the original number of vertices.

Our initial goal was to determine how template generation would affect the general scheduling of instructions, regardless of whether the target of compilation is a super-scalar pipelined architecture or the synthesis of new hardware. Although application synthesis is not the goal of the compiler, a high-level synthesis tool could perform scheduling of the resulting clustered DFGs. Specifically, our experiment is designed to determine the scheduling latency of our generated DFGs (intuitively analogous to the time of execution on a powerful processor). We compare this latency to the scheduling latency of the original (nonclustered) DFG. Assuming that latency is generally improved by the addition of special blocks of logic to execute regular instructions, we furthermore wish to explore the impact that these clustering decisions will make on chip area. Specifically, a thorough exploration of the latency/area tradeoffs of clustering is required in order to evaluate our methods.

In order to simulate the results of our algorithm, we compiled and generated instructions for four programs: an image convolution algorithm [274], DeCSS (the decryption of DVD encoding) [153], the DES encryption algorithm [275], and the Rijndael AES encryption algorithm [276]. These algorithms are typical candidates for industrial hardware implementation (as cameras, DVD players, and embedded encryption devices must perform these operations). Additionally they are computationally intensive, leading to generally large DFGs which are benign to regularity extraction. From each compiled CDFG of the programs, four representative DFGs were selected for scheduling. The scheduling algorithm we used has been described in detail in [261], and is comparable to the state-of-the-art in the research community. The resulting latency of each scheduled DFG (both with and without clustering) is recorded in Figure 51. In Figure 52, we record both the decrease in latency and the increase in FPGA area that resulted from our clustering algorithm.

		No Clustering	Parallel And Sequential Clustering
Convolve	Node 1	10	5
	Node 2	6	4
	Node 3	8	2
	Node 4	8	6
DeCSS	Node 1	11	6
	Node 2	10	3
	Node 3	56	31
	Node 4	21	9
DES	Node 1	84	20
	Node 2	59	24
	Node 3	20	11
	Node 4	18	11
Rijndael AES	Node 1	24	15
	Node 2	56	32
	Node 3	17	6
	Node 4	6	2

Figure 51 Latency measurements for each scheduled DFG (in clock cycles)

Clearly, clustering of DFGs reduces the number of total instructions, and increases the potential to execute frequently occurring sets of parallel operations. This directly improves the schedule of the application DFGs, demonstrating latency improvement by as much as 76.19% on our largest DFG (the first basic block of the DES encryption algorithm: 150 nodes). For every DFG scheduled, latency was improved by at least 25%, a surprisingly good figure. Additionally, the FPGA area increased an average of 21.55% (maximally 150% in some smaller DFGs). Occasionally, even decreased area was realized via clustering, presumably due to improved utilization of regular specialized components. Overall, the average latency improvement (51.98%) shadowed the area gains (average 21.55%), especially on larger, more complex DFGs.

		Size of original DFG (nodes)	% Latency Decrease	% FPGA Area Increase
Convolve	1	20	50	66.67
	2	13	33.33	-4.55
	3	17	75	31.25
	4	19	25	4.17
DeCSS	1	21	45.45	150
	2	13	70	-12.5
	3	121	44.64	25
	4	55	57.14	37.5
DES	1	150	76.19	-5.83
	2	122	59.32	23.95
	3	55	45	15.38
	4	43	36.89	4.17
Rijndael AES	1	38	37.5	20.91
	2	105	42.86	33.00
	3	46	64.71	-6.25
	4	8	66.67	-38.10
averages:		52.875	51.98	21.55

Figure 52 Latency/Area Impact of Clustering

8. RELATED WORK

While there has been a lot of work on regularity extraction, most of it focuses on template matching (similar to the graph covering problem) and not template generation. Regularity extraction was shown beneficial in reducing area and increasing performance for the PipeRench architecture [251]; PipeRench is a fully reconfigurable, pipelined FPGA. The benefits stem from the fact that the templates may be hand optimized. Additionally, the template operations are placed in the same vicinity on the chip. This reduces the interconnect delay as well as compacts the application into a smaller portion of the chip. Cadambi and Goldstein proceed to show that templates lead to a decrease in area and delay for the PipeRench architecture; they suggest that profiling is beneficial for small granularity FPGAs e.g. LUT or PLA-based, though no empirical evidence is given to support this claim.

Cadambi and Goldstein restrict their template generation to single output templates and limit the number of inputs. If the templates are going to be used as soft reconfigurable macros and placed in a configurable fabric, the number of inputs/outputs must be limited to maintain good routability. But, templates can also be used to generate the VPB functionality. Since the VPBs are ASIC blocks integrated into the reconfigurable fabric, the system architecture can place additional routing resources around VPBs to handle the additional routing needed by VPBs with a large number of inputs/outputs. Therefore, generated templates need not always have input/output restrictions.

Regularity extraction is used in a variety of other CAD applications. Templates are used during scheduling to address timing constraints and hierarchical scheduling [277]. Data path circuits exhibit a high amount of regularity; hence regularity extraction reduces the complexity of the program as well as increasing the quality of the result [278, 279]. System level partitioning is yet another use of regularity extraction [266]. Furthermore, proper use of templates can lead to low-power designs [280].

One of the earliest template matching works in the CAD community was by Kahrs [281], wherein a greedy, bottom-up procedure for a silicon compiler is described. Keutzer [282] modeled a system as a DAG and heuristically partitioned it to yield rooted trees and applied compiler techniques to test for pattern matches. Trees and single output templates are used by Chowdhary et al. [279] to cover data path circuits.

Rao and Kurdahi [266] addressed template generation for system-level clustering using the well-known first fit approach to bin filling. More recently, Cadambi and Goldstein [265] propose single output template generation via a constructive, bottom up approach. Both methods restrict the area and the number of pins for their templates. Our method attempts to find the best possible set of templates, regardless of area and size, though we can easily add pin and area restrictions to our algorithms. Additionally, we perform template generation and matching simultaneously.

IMEC's Cathedral Project [283] used a different model of computation in their high-level synthesis stage: instead of a CDFG they performed reductions on the signal flow graph of a DSP application. Their data path was composed of Abstract Building Blocks (ABBs), or instructions available from a given hardware library. The customized data path generated from many ABBs was referred to as an application specific unit (ASU). Cathedral's synthesis targeted ASUs, which could be executed in very few clock cycles. This goal was achieved via manual clustering of necessary operations into more compact operations, essentially a form of template construction. Whereas our template generation and matching algorithms are automated, the definition of clusters in Cathedral was a manual operation, mainly clustering loop and function bodies. Their results demonstrated an expected reduction of critical path length as well as interconnect as a result of clustering.

One of the more encouraging cases of performance gain via template matching was investigated by Corazao et al [284]. Their work assumed a given library of highly regular templates. These templates could be utilized during the high-level synthesis stage in order to minimize the number of clock cycles in a circuit's critical path. In circumstances where some parts of a template were not needed, partial matching was also allowed. With partial matching, some portions of a selected template go unused. Their experimental results demonstrated large performance gains without an unreasonable increase in area. Although many optimization techniques were utilized as part of the synthesis strategy, template selection had the largest impact on overall improvement in throughput.

The Totem Project [285, 286] endeavors to automate the generation of custom reconfigurable architectures based on a given set of applications. Built upon the RaPiD architecture [252], their optimizations are made at the placement and routing stages of synthesis, mapping coarse-grained components to a one-dimensional data path axis. Unlike our design, their input is a set of architecture netlists, which are transformed directly to a physical design while targeting the simultaneous goals of increased routing flexibility and decreased area.

9.   SUMMARY

In this work, we addressed the problem of instruction generation. We discussed two algorithms to solve the problem which performs simultaneous template generation and matching: the Sequential Template Generation Algorithm and its successor, the Parallel Template Generation Algorithm. These algorithms generate instructions by profiling a dataflow graph and clustering common edges. Furthermore, we present some theory behind instruction generation, including the complexity of these solutions.

Instruction generation is a relatively new and essential problem for compilation to reconfigurable systems. Instruction generation can be used to create soft reconfigurable macros, which are tightly coupled sequential operations that are placed in the same vicinity in a configurable fabric. Furthermore, the macros are ideal candidates for hand optimization. Additionally, template generation can be used to specify the functionality for pre-placed ASIC blocks (VPBs) in hybrid reconfigurable systems.

We developed a co-compiler for a hybrid reconfigurable system, and have incorporated the discussed template generation techniques. Our latest results demonstrate large performance gains on computationally intensive algorithms with minimal FPGA area increases.

In the future, we intend to parameterize our algorithm, enabling it to produce different instructions given different system constraints (such as maximum area increase). This will allow a new level of hardware awareness in our compiler, allowing it to improve latency only when it is physically reasonable to do so. Additionally, we wish to explore DFG inlining and translations to other intermediate representations, which will provide us with a more global view of the application, as well as help us to determine other forms of instruction regularity.

Chapter 8

DATA COMMUNICATION

In order to facilitate the future design of embedded systems and system-on-chips, we must develop techniques to explore the design space of the system. These tools will take a high-level specification of the application and produce a customized hardware system. The system may be comprised of many components – different types of processors (e.g. ARM, VLIW, superscalar) as well as reconfigurable logic devices (e.g. FPGA) and/or ASIC components.

The compiler straddles the boundary between application and hardware, making it a natural area to perform system exploration. The compiler can already map portions of the application to different processors by simply emitting code. This only allows exploration on a system composed of various numbers and types of processors. In order to complete the system exploration space – one with processors, ASIC and reconfigurable components, we need a path from the compiler to a hardware description language (HDL); this allows us to map portions of the application to ASICs, FPGAs and any other devices that accept HDL as an input.

An area of extreme importance is the translation of the compiler's intermediate representation (IR) to a form that is suitable for synthesis to hardware. During this translation, we should attempt to exploit the existing concurrency of the application and discover additional parallelism [117]. Also, we should determine the types of hardware specialization that will increase the efficiency of the application [98, 287]. Finally, we must take into account the hardware properties of the circuit, e.g. power dissipation, critical path and interconnect area.

Static single assignment [288, 289] transforms the IR such that each variable is defined exactly once. It is an ideal transformation for hardware because lone side effects of the transformation, Φ-nodes, are easily implemented in hardware as multiplexers. It has been used in many projects where the final output is an HDL [29, 109, 218]. Yet, SSA was originally developed to enable optimizations for fixed architectures; it was not originally meant for hardware synthesis.

In this work, we study SSA and its effect on the optimization of hardware properties of the circuit. We show how SSA can be used to minimize data communication; this has a direct effect on the area, amount of interconnect and delay of the final circuit. Furthermore, we show that SSA in its original form is not optimal in terms of data communication and give an optimal algorithm for the placement of Φ-nodes to minimize the amount of data communication.

In the next section, we give background material related to our research. We show how SSA is useful to minimize interconnect in the hardware in Section 1. Furthermore, we point out a fundamental shortcoming of traditional SSA and develop a new SSA algorithm to overcome this limitation. Section 2 presents experiments to illustrate the effect of these algorithms to minimize data communication. Section 3 discusses a refinement of our SSA algorithm, utilizing information from circuit floorplans to recompile the hardware. Section 4 presents experiments using this refined approach. We discuss related work in Section 5 and provide concluding remarks in Section 6.

1.   MINIMIZING INTER-NODE COMMUNICATION

In order to determine the data exchange between nodes in a CDFG, we must establish the relationship between where data is generated and where data is used for calculation. The specific place where data is generated is called its *definition point*, and the place where that data is used in computation is called a *use point*. The data generated at a particular definition point may be used in multiple places. Likewise, particular use point may correspond to a number of different definition points; the control flow may dictate the actual definition point used in later computation.

In hardware, the path of execution between nodes can be controlled either by a central controller or by the nodes themselves. The former is called centralized control, whereas the latter is decentralized control. The hardware nodes are referred to as control nodes, and roughly correspond to nodes on a compiler's control flow graph.

If data generated in one control node is used in a computation in a second control node, these two control nodes must have a mechanism to transfer the data between them. A distributed data communication scheme has a direct connection between the two control nodes (i.e. one node controls the other's execution through a signal). If a centralized scheme were used, the first control node would transfer the data to memory and the second control node would access the memory for that data. Therefore, in a centralized scheme minimizing the inter-node communication would have a direct impact on the number of memory accesses, and in a distributed scheme the interconnect between the control nodes would be reduced. However, in both control schemes real performance boosts can be realized through communication optimization. Thus, regardless of the scheme that we use, we should try to minimize the amount of inter-node communication.

1.1　Static Single Assignment

We can determine the relationship between the use and definition points through a compiler intermediate form known as static single assignment [288, 289]. Static single assignment (SSA) renames variables with multiple definitions into distinct variables – one for each definition point. A variable in a high-level programming language has a *name*, which represents the contents of a storage location (e.g. register, memory). A name is not specific to SSA. In non-SSA code, a name also represents a value residing in a storage location but we may not yet know the exact location or the value that the name represents, as these attributes sometimes depend on the control flow of the program. Therefore, we refer to a name in non-SSA code as a *location*. SSA eliminates this confusion as each name represents a value that is generated at exactly one definition point. In SSA form, a definition (or assignment) maps a single value to a single name, and that name represents that value for the entire program.

In order to maintain proper program functionality, we must add Φ-nodes into the CDFG. Φ-nodes are needed when a particular use of a name is defined at multiple points. A Φ-node takes a set of possible names and outputs the correct one depending on the path of execution. Φ-nodes can be viewed as an operation of the control node. They can be implemented using a multiplexer. Figure 53 illustrates the conversion to SSA.

Transformation of a CDFG into SSA form is accomplished in two steps, first we add Φ-nodes and then we rename the variables at their definition and use points. There are several methods for determining the location of the Φ-nodes. The naïve algorithm would insert a Φ-node at each merging point for each original name used in the CDFG. A more intelligent algorithm – called the minimal algorithm – inserts a Φ-node at the *iterated dominance frontier* of each original name [288]. The iterated dominance frontier is the set of control nodes in the timeline of the program at which two or more definitions of a variable merge. The semi-pruned algorithm builds smaller SSA form than the minimal algorithm. It calculates

Data Communication

determines if a variable is local to a basic block and only inserts Φ-nodes for non-local variables [289]. The pruned algorithm further reduces the number of Φ-nodes by only inserting Φ-nodes at the iterated dominance frontier of variables that are live at that time [290]. After the position of the Φ-nodes is determined, there is a pass where the variables are renamed.

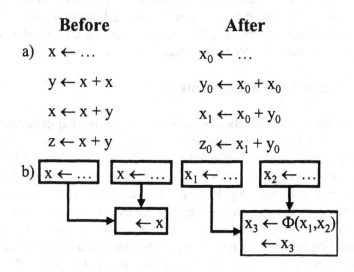

Figure 53 a) Conversion of straight-line code to SSA b) SSA conversion with control flow

The minimal method requires $O(|E_{cfg}| + |N_{cfg}|^2)$ time for the calculation of the iterated dominance frontier. The iterated dominance frontier and liveness analysis must be computed during the pruned algorithm. There are linear or near linear time liveness analysis algorithms [291-293]. Therefore, the pruned method has the same asymptotic runtime as the minimal method.

We should suppress any unnecessary data communication between control nodes. Now we explain how to minimize the inter-node communication.

1.2   Minimizing Data Communication with SSA

SSA allows us to minimize the inter-node communication. The various algorithms used to create SSA all attempt to accurately model the actual need for data communication between the control nodes. For example, if we use the pruned algorithm for SSA, we eliminate false data communication by using liveness analysis, which eliminates passing data that will never be used again.

SSA allows us to minimize the data communication, but it introduces Φ-nodes to the graph. We must add a mechanism that handles the Φ-nodes. This can be accomplished by adding an operation that implements the functionality of a Φ-node. A multiplexer provides the needed functionality. The input names are the inputs to the multiplexer. An additional control line must be added for each multiplexer to determine that the correct input name is selected.

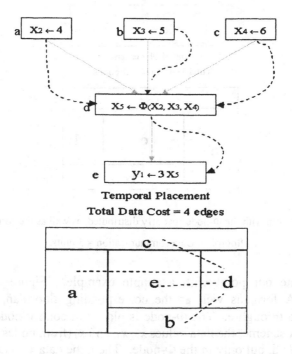

Figure 54 SSA form and the corresponding floorplan (dotted edges represent data communication, and grey edges represent control). Data communication = 4 units.

A fundamental limitation of using SSA in a hardware compiler is the use of the iterated dominance frontier for determining the positioning of the Φ-nodes. Typically, compilers use SSA for its property of a single definition point. We are using it in another way – as a representation to minimize the data communication between hardware components (CFG nodes). In this case, the positioning of Φ-nodes at the iterated dominance frontier does not always optimize the data communication. We must consider spatial properties in addition to the temporal properties of the CDFG when determining the position of the Φ-nodes.

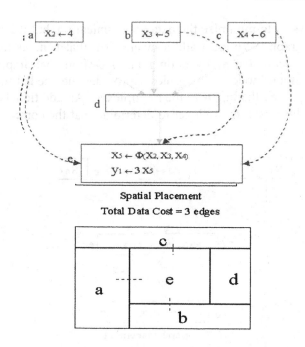

Figure 55 SSA form with the Φ-node spatially distributed, as well as the corresponding

floorplan. Data communication = 3 units.

We illustrate our point with a simple example. Figure 54 exhibits traditional SSA form as well as the corresponding floorplan, containing control nodes a through e. The Φ-node is placed in control node d. In the traditional SSA scheme, the data values x_2, x_3, and x_4 (from nodes a, b, and c) are used in node d, but only in the Φ-node. Then, the data x_5 is used in node e. Therefore, there must be a communication connection from node a to node d, node b to node d and node c to node d, as well as a connection from node d to node e – a total of 4 communication links. In Figure 55, the Φ-node is distributed to node e. Then, we only need a communication connection from nodes a,b, and c to node e, a total of 3 communication links.

From this example, we can see that traditional Φ-node placement is not always optimal in terms of data communication. This arises because Φ-nodes are traditionally placed in a temporal manner. To intuit why Φ-nodes are placed as temporally early as possible, note that a Φ-node is the point at which all of its source's live ranges end. For example, consider the assignment $x_5 = \Phi(x_2, x_3, x_4)$. From a temporal perspective (i.e. looking at the control flow statically, without considering loop behavior), variables x_2, x_3, and x_4 are no longer live after this Φ-assignment statement. In other words, no registers need to be reserved for these three variables beyond this point. Therefore, by placing Φ-nodes temporally, live ranges are kept as short as possible, and register contention is reduced. The iterated dominance frontier is the first place in the timeline of the program where the two (or more) locations of a variable merge. Thus, in software compilation, the iterated dominance frontier (or temporal placement location) is the best location for Φ-nodes to be placed.

When considering hardware compilation, we must think spatially as well as temporally. Put differently, the iterated dominance frontier is not the only place where a Φ-node can be positioned, and in the case of hardware, it may not be the best location. By moving the position of the Φ-nodes, it is possible to achieve a better layout of our hardware design. In order to reduce the data communication, we must consider the number of uses of the value that a Φ-node defines as well as the number of values that the Φ-node takes as an input.

1.3 An Algorithm for Spatially Distributing Φ-nodes

The first step of spatially distributing Φ-nodes is determining which Φ-nodes should be moved. We assume that we are given the correct temporal positioning of the Φ-nodes according to some SSA algorithm (e.g. minimal, semi-pruned, pruned). The movement of a Φ-node depends on two factors. The first factor is the number of values that the Φ-node must choose between. We call this number of Φ-node source values s. The second factor is the number of uses that the value of the Φ-node defines. We call this number of uses of the Φ-node d. Taking Figure 54 as an example, the Φ-node source values are x_2, x_3, and x_4 whereas the Φ-node destination value is x_5. Determining s is simple; we just need to count the number of source values in the Φ-node. Finding the number of uses of the destination value is a more difficult. We can use def-use chains [294], which can be calculated during SSA.

The relationship between the amount of communication links C_T needed for a Φ-node in temporal SSA and the number of communication links C_S in spatial SSA is:

$$C_T = s + a \qquad C_S = s \cdot a$$

Using these relationships, we can easily determine if spatially moving a Φ-node will decrease the total amount of inter-node data communication. If C_S is less than C_T, then moving the Φ-node is beneficial. Otherwise, we should keep the Φ-node in its current location. After we have decided on which Φ-nodes we should move, we must determine the control node(s) where we should move the Φ-node. This step is rather easy, as we move the Φ-node from its original location to control nodes that have a use of the definition value of that Φ-node.

```
1.  Given a CDFG G(N_cfg, E_cfg)
2.  perform_SSA(G)
3.  calculate_def_use_chains(G)
4.  remove_back_edges(G)
5.  topological_sort(G)
6.  for each node n ∈ N_cfg
7.      for each Φ-node Φ ∈ n
8.          s ← |Φ.sources |
9.          d ← |def_use_chain(Φ.dest)|
10.         if s · d < s + d
11.             move_to_spatial_locations(Φ)
12. restore_back_edges(G)
```

Figure 56 Spatial SSA algorithm

It is possible that a use point of the definition value of Φ-node Φ_1 is another Φ-node Φ_2. If we wish to move Φ_1, we add the source values of Φ_1 into the source values of Φ_2; obviously, this action changes the number of source values of Φ_2. In order to account for such changes in source values, we must consider moving the Φ-nodes in a topologically sorted manner based on the CDFG control edges. Of course, any back control edges must be removed in order to have valid topologically sorting. We can not move Φ-nodes across back edges as this can induce dependencies between the source value and the destination value of previous iterations i.e. we can get a situation where $b_1 \leftarrow \Phi(b_1, ...)$. The source value b_1 was produced in a previous iteration by that same Φ-node. The complete algorithm for spatially distributing Φ-node to minimize data communication is outlined in Figure 56.

Theorem 8.1: Given an initially correct placement of a Φ-node, the functionality of the program remains valid after moving the Φ-node to the basic block(s) of all the use point(s) of the Φ-node's destination value.

Proof: There are two cases to consider. The first case is when the use point is a normal computation. The second case is when a use point is Φ-node itself.

We consider the former case first. When we move the Φ-node from it's initial basic block, we move it to the basic blocks of every use point of the Φ-node's destination value d. Therefore, every use of the d can still choose from the same source values. Hence, if the Φ-node source values where initially correct, the use points of d remain the same after the movement. We must also insure that moving the Φ-node does not cause some other use point that uses the same name but has a different value. The Φ-node will not move past another Φ-node that has the same name because by construction of correct initial SSA, that Φ-node must have d as one of its source values.

The proof of the second case follows similar lines to that of the first one. The only difference is that instead of moving the initial Φ-node Φ_i to that basic block, we add the source values to the Φ-node Φ_u that uses d. If we move Φ_i before Φ_u, then the functionality of the program is correct by the same reasoning of the first part of proof. Assuming that the temporal SSA algorithm has only one Φ-node per basic block per name, we can add the source values of Φ_i to Φ_u while maintain the correct program functionality. □

Theorem 8.2: Given a correct initial placement of Φ-nodes, the spatial SSA algorithm maintains the correct functionality of the program.

Proof: The algorithm considers the Φ-nodes in a topologically sorted manner. As a consequence of Theorem 6.1, the movement of a single Φ-node will not disturb the functionality of the program hence the Φ-node will not move past another value definition point with the same name. Since we are considering the Φ-nodes in forward topologically sorted order, the movement of any Φ-node will never move past a Φ-node which has yet to be considered for movement. Also, a Φ-node can never move backwards across an edge (remember we remove back edges). Therefore, the algorithm will never move a value definition point past another value definition point with the same name. Hence every use preserves the same definition after the algorithm completes. This maintains the functionality of the program. □

Theorem 8.3: Given a floorplan where all wire lengths are unit length, the Spatial SSA Algorithm provides minimal data communication.

Proof: The source values of any given Φ-function are individual control nodes, and the cardinality of these nodes shall be referred to as s. Likewise, the destination points of any Φ-function are individual control nodes, and their cardinality will be referred to as d. The number of control nodes which define a given Φ-function will be referred to as n. The amount of data communication that this algorithm can reduce is restricted to the number of data edges coming into each Φ-node and the number of data edges coming out of each Φ-node. (The other data communication is already minimized, since SSA variables are actual data values. Therefore, SSA variables passed between control blocks are actual pieces of data that must be moved.) If a Φ-function is spatially moved to one of its use points, then the number of out degree edges specifically leaving this Φ-node can be considered equal to zero. (The Φ-node's out degree data edges are now equal to the out degree of the use point, which cannot be reduced any further by the placement or removal of the Φ-node. Therefore the Φ-node's out degree of data will be considered equal to zero in this case.)

The total number of data communication points entering and exiting the Φ-nodes of a given Φ-function can be represented by a cost equation:

$$C = \sum_{n\,\Phi\,nodes} in + out$$

where *in* is the number of inbound edges to each Φ-node and *out* is the number of outbound edges from each Φ-node.

In a floorplan where each edge has unit cost, this equation represents the total cost of this Φ-function in the graph.

In order to maintain correctness in a CDFG, every source value of a Φ-function must be coming into all Φ-nodes defining this function. (This is the only data that needs to enter a Φ-node.) Therefore, for all minimal cost cases, we can say that $in = s$ for every Φ-node and the data communication cost of the Φ-function can be restated as $C = ns + \sum_{n\,\Phi\,nodes} out$ since s is constant.

This leaves us with two values we can minimize: *n* (the number of total nodes defining a given Φ-function) and *out* (the out degree of a Φ-node), since *s* cannot be reduced (for correctness's sake). The most minimal cost we can have is when $n = 1$ or $out = 0$. ($n >= 1$, because at least one node must define the Φ-function. $out = 0$ is possible, as stated earlier.)

In the case that $out = 0$, the Φ-function will be coalesced with every use point of that function. That means that the total number *n* of nodes defining this function will equal *d* (the number of use points of the Φ-function). Therefore, $C = ns + \sum_{n\,\Phi\,nodes} out = ns = d \cdot s = \mathbf{s} \cdot \mathbf{d}$ (corresponding to spatial placement)

In the case that $n = 1$, that means that there is only one node defining a given Φ- function. This means that either a) there is a directed edge from this node to every use point or b) there is only one use point and this node has been coalesced with it.

In the case of part a, the total number of directed edges leaving the one Φ-node is equal to *d* (the number of use points) therefore $C = 1 * s + \sum_{n\,\Phi\,nodes} out = s + out = \mathbf{s} + \mathbf{d}$ (corresponding to temporal placement)

Part b is a special case of $C = s * d$ ($n = 1$, *out* = 0). Therefore, we can minimize the total in/out degree of the Φ-node(s) by minimizing the equations ($C = s + d$, $C = s * d$). This corresponds to either choosing temporal placement (in the case of $s + d < s * d$) or choosing spatial placement (if $s + d > s * d$). This minimization of the degree of the Φ-node(s) leads to minimal data communication in the CDFG. □

2.   INITIAL EXPERIMENTAL RESULTS

To measure the effectiveness of using SSA to minimize data communication between control nodes, we examined a set of DSP functions. DSP functions typically exhibit a large amount of parallelism making them ideal for hardware implementation. The DSP functions were taken from the MediaBench test suite [188] (See Table 3). The files were compiled into CDFGs using the SUIF compiler infrastructure [165] and the Machine-SUIF [166] backend.

We performed SSA analysis with the SSA library built into Machine-SUIF. The library was initially developed at Rice [295] and recently adapted into the Machine-SUIF compiler.

First, we compare the amount of data communicated between the control nodes using the different SSA algorithms. Given two control nodes i and j, the *edge weight* $w(i,j)$ is the amount of data communicated (in bits) from control node i to control node j. The *total edge weight (TEW)* is:

$$TEW = \sum_i \sum_j w(i, j)$$

Figure 57 is a comparison of edge weights using three different algorithms for positioning the Φ-nodes. We compare the minimal, semi-pruned and pruned algorithms. Recall that the pruned algorithm is the best algorithm in terms of reducing the number of Φ-nodes, but worst in runtime. The minimal algorithm produces many Φ-nodes, but has small runtime. The semi-pruned algorithm provides a middle ground in terms of runtime and quality of result.

We divide the TEW of the minimal and semi-pruned algorithm (respectively) by the TEW of the pruned algorithm. We call this the *TEW ratio*. We use the pruned algorithm as a baseline because it consistently produces the smallest TEW. Referring to Figure 57, the TEW of the minimal algorithm is much worse than that of the pruned algorithm. For example, in the benchmark fft2, the TEW of the minimal algorithm is over 70 times that of the TEW of the pruned algorithm. The semi-pruned algorithm yields a TEW that is smaller than that of the minimal algorithm, but still slightly larger than the TEW of the pruned algorithm. All algorithms have the same asymptotic runtime and the actual runtimes for all the algorithms over all the benchmarks were very small (under 1 second). Therefore, we feel that one should use the pruned algorithm as it minimizes data communication much better than the other two algorithms. Furthermore, the actual additional runtime needed to run the pruned algorithm is miniscule.

Table 3 MediaBench functions

Application	C File	Description
mpeg2	getblk.c	DCT block decoding
adpcm	adpcm.c	ADPCM to/from 16-bit PCM
epic	convolve.c	2D general image convolution
jpeg	jctrans.c	Transcoding compression
rasta	fft.c	Fast Fourier Transform
rasta	noise_est.c	Noise estimation functions

Function	Name	# Control Nodes
adpcm_coder	adpcm1	33
adpcm_decoder	adpcm2	26
internal_expland	convolve1	101
internal_filter	convolve2	101
compress_output	jctrans	33
decode_MPEG2_intra_block	getblk1	75
decode_MPEG2_non_intra_block	getblk2	60
decode_motion_vector	motion	15
FAST	fft1	14
FR4TR	fft2	76
comp_noisepower	noise_est1	153
Det	noise_est2	12

Each of the algorithms we compared attempt to minimize the number of Φ-nodes, and not the data communication. There is obviously a relationship between the number of Φ-nodes and the amount of data communication. Every Φ-node defines additional data communication, but there can be inter-node data transfer without Φ-nodes. Furthermore, as we pointed out in Section 1, minimizing the number of Φ-nodes does not directly correspond to minimizing the data communication.

In Figure 59, we compare the ratio of Φ-nodes and the ratio of TEW using the minimal and pruned algorithms. As you can see, the number of Φ-

nodes is highly related to the amount data communication. As the Φ-node ratio increases, the TEW ratio increases. Correspondingly, a large Φ-node ratio corresponds to a large TEW ratio. This lends validation to the using SSA algorithms to first minimize inter-node communication and then use the spatial Φ-node repositioning to further reduce the data communication. In other words, minimizing the number of Φ-nodes is a good objective function to initially minimize data communication.

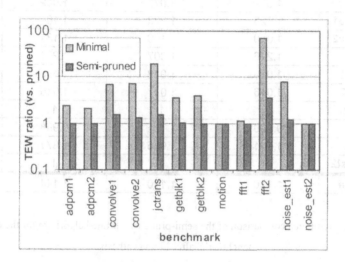

Figure 57 Comparison of total edge weight (TEW) between the minimal and semi-pruned TEW and the pruned TEW

We have focused on the total edge weight as a model of "goodness" for realizing the circuit in hardware. This model makes intuitive sense; by minimizing the amount of data that we pass within the circuit we should produce a "good" implementation of that circuit. Essentially, we are arguing that the TEW is a good metric in which to judge the quality of a circuit implementation.

In order to judge TEW as a metric of the quality of a circuit implementation, we used our system synthesis framework (described in Chapter 5) to realize each of the benchmarks as a hardware implementation. We synthesized each of the benchmarks using Synopsys Behavioral Compiler for architectural synthesis followed by Synopsys Design Compiler for logic synthesis. Then, we gathered the area statistics for each of the designs using the three different SSA algorithms. We present these statistics in Table 4 and Table 5.

Data Communication

Table 4 A circuit area comparison of the semi-pruned vs. pruned algorithm for the different benchmarks after logic synthesis.

Benchmark	Combinatorial Area	Non-combinatorial Area	Net Interconnect Area	Total Area
Adpcm1	1.066	1.100	1.074	1.081
Adpcm2	1.037	1.073	1.049	1.054
Convolve1	1.123	1.223	1.168	1.164
Convolve2	1.010	1.169	1.134	1.129
Jctrans	1.226	1.307	1.271	1.267
Getblk1	1.006	1.013	1.005	1.008
Getblk2	1.016	1.017	1.016	1.016
Motion	1	1	1	1
Fft1	0.960	1	0.998	0.987
Fft2	1.741	1.965	1.871	1.851
Noise_est2	1	1	1	1
Average	1.116	1.170	1.144	1.142

Table 5 A circuit area comparison of the semi-pruned vs. pruned algorithm for the different benchmarks after logic synthesis.

Benchmark	Combinatorial Area	Non-Combinatorial Area	Net Interconnect Area	Total Area
Adpcm1	1.833	1.969	1.941	1.918
Adpcm2	1.880	2.067	1.985	1.982
Getblk1	2.933	4.636	3.096	3.461
Motion	1	1	1	1
Fft1	2.376	1.218	2.105	1.867
Noise_est2	1	1	1	1
Average	1.837	1.982	1.854	1.871

The two tables show three different elements for the area of the circuit. The combinatorial area is the area used by the gates that implement data path components (e.g. adders, multipliers). The non-combinatorial area is that of the memory and steering logic components (e.g. flip-flops, multiplexers). The area of the wires is given by the net interconnect area. The total area is simply the sum of the three previous elements. We give the ratio of the area of the semi-pruned (minimal) algorithm versus the pruned algorithm in a similar manner as the TEW ratios. Some of the benchmarks were omitted from the tables. These benchmarks ran out of memory on our servers (even a 4 processor Sun server with 2 GB memory!)

Figure 58 charts the total area ratios of the benchmarks. The figure demonstrates that our assumptions about minimizing the TEW have a good correlation with minimizing the area of the circuit. Comparing Figure 57 with Figure 58, you can see that the amount of reduction in TEW correlates with the amount of reduction in total area. For example, the TEW for the benchmark fft2 using the semi-pruned algorithm is approximately 5 times that of the pruned algorithm. A similar result is seen in the total area ratio; the area of the semi-pruned algorithm is about 1.8 times that of the area of the pruned algorithm. Furthermore, the area of the pruned algorithm is almost always the best algorithm in terms of total area. Benchmark fft1 is the lone exception, most likely due to its small size and limited number of Φ-nodes. And even with this lone outlier, the overall average area improvements are 14% and 87% better using the pruned algorithm over the semi-pruned and minimal algorithm. It is safe to say that the type of SSA algorithm has a huge effect on the area of the circuit implementation. Furthermore, the results indicate that it is worth the small increase in runtime to use the pruned algorithm.

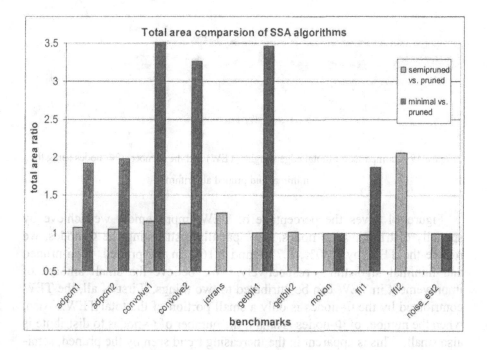

Figure 58 A total area comparison of the benchmarks after logic synthesis. The ratio is the minimal (semipruned) total area divided by the pruned total area.

Our next set of experiments focus on using spatial SSA Φ-node distribution to further minimize the amount of data communication. Figure 60 shows the number of Φ-nodes that are spatially distributed by the spatial SSA algorithm. We can see that these Φ-nodes are fairly common; in some of the benchmarks, over 35% of the Φ-nodes are spatially moved. The average number of distributed Φ-nodes over all the benchmarks is 11.65%, 18.21% and 13.56% for the pruned, semi-pruned and minimal algorithms, respectively. Not all of the benchmarks are included in Figure 60; the omitted benchmarks have 0 Φ-nodes that should be distributed, but these benchmarks are included in the averages.

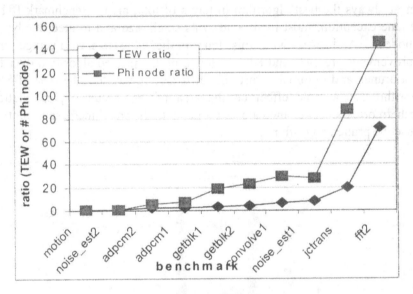

Figure 59 A comparison of total edge weight (TEW) and the number of Φ-nodes using the

minimal and pruned algorithms.

Figure 61 gives the percentage of TEW improvement we achieve by spatially distributing the nodes. By spatially distributing the Φ-nodes, we reduce the TEW by 1.80%, 4.77% and 8.16% in the pruned, semi-pruned and minimal algorithms, respectively. We believe the small amount of improvement in TEW can be attributed to two things. First of all, the TEW contributed by the Φ-nodes is only a small portion of the total TEW. Also, when the number of Φ-nodes is small, the number of Φ-nodes to distribute is also small. This is apparent in the increasing trend seen by the pruned, semi-pruned and minimal algorithms. There are many Φ-nodes when we use the minimal algorithm and correspondingly, there TEW improvement of the minimal algorithm is 8.16%. Conversely, the number of Φ-nodes in the pruned algorithm is small and the TEW improvement is also small.

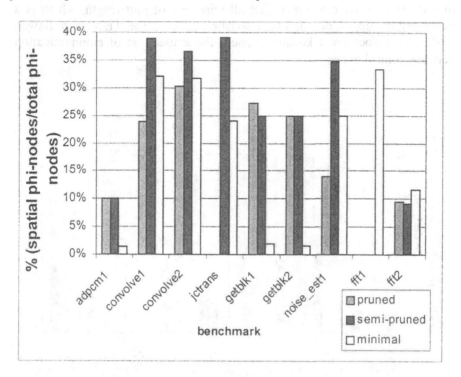

Figure 60 Comparison of the number of spatially distributed Φ-nodes and the total number Φ-nodes using the three SSA algorithms.

We ran the spatial algorithm through our system framework to determine the actual area improvements achieved by spatially distributing the Φ-nodes. The results are shown in Figure 62. The results are mostly negative. The chart plots the total area of the temporal (original) Φ-node placement divided by the total area of the spatial Φ-node placement as described our spatial Φ-node placement algorithm. A result above 1 denotes that the temporal area is larger than the spatial area, meaning that our spatial Φ-node placement algorithm is beneficial. Sadly, almost no bar is above 1. In other words, there is hardly ever an area reduction resulting from spatial placement, and when present this reduction is very slight. The other benchmarks either have worse total area or the total area is approximately the same i.e. the total area ratio is equal to 1.

We believe that these initial results are negative for two reasons. First, as stated previously, the TEW reduction when using the spatial algorithm is not that large. The TEW reduction was 1.80%, 4.77% and 8.16% using the pruned, semi-pruned and minimal algorithms.   Second, and more importantly, we have assumed that all wires are of unit length, which is a naïve estimation of circuit characteristics.   Thus, the TEW is a flawed model, as it does not take into account the actual cost of communication between control nodes.

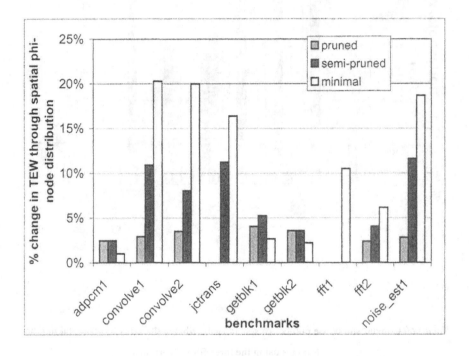

Figure 61 The percentage change in of total edge weight when we distribute the Φ-nodes using the three SSA algorithms.

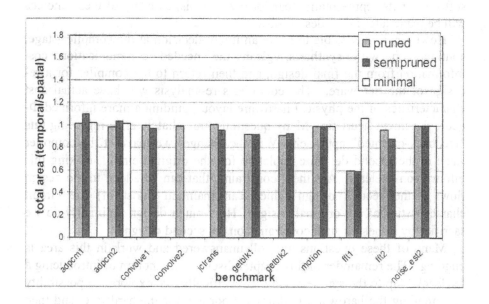

Figure 62 Comparison of the total area of the temporal versus spatial Φ-node placement for

the three SSA algorithms.

3. DESIGN FEEDBACK LOOP

Minimization of TEW is a simple objective which can be achieved by a compiler using SSA form. However, meeting this objective buys us little in terms of actual synthesized hardware area. Returning to the example in Figure 54 and Figure 55, we know that spatial placement of Φ-nodes may be beneficial to data communication and wirelength in the final synthesized circuit. In general, a Φ-node may be placed temporally (at the iterated dominance frontier), spatially (at each of its uses), or even somewhere in between, as long as program correctness is maintained. Furthermore, a Φ-node may be replicated, which in some cases may add more wires but result in reduced total wirelength. To effectively explore these design possibilities, we need to create a more informed compiler-level heuristic. Specifically, the compiler needs to understand the actual cost of communicating (sending data) between one part of a program and another. A simple analysis of the software or its representation cannot give us this, as many of these choices will be made later in the design flow.

However, it is possible to make an initial decision at the compiler stage, continue on with synthesis, glean the needed communication cost information from the final design, and then return to the compiler for a re-design of the hardware. The compiler's re-analysis will have actual cost characteristics of the physical hardware layout, guiding a more informed Φ-node placement heuristic. Many questions immediately arise regarding this "feedback-oriented" approach. Will the changes we make at the compiler stage in the second design completely foil the original circuit, making this information useless? Are there constraints that can be added to the design flow in the second design which can maintain (most of) the circuit characteristics of the original design? How much design optimality do we sacrifice by imposing such constraints on the second design?

Many of these questions are still unanswered and work in this area is ongoing. The remainder of this chapter discusses the notion of introducing a feedback loop to the synthesis flow. Specifically, we explore the possibility of compiling the hardware, building a floorplan for the hardware, and then using the floorplan to obtain cost information (i.e. the cost of moving data between pieces of the program). We then re-compile the hardware with the original floorplan's cost information, and heuristically re-distribute Φ-nodes. This should result in improved placement of Φ-nodes for reduced data communication.

3.1   Intuition Behind the New Problem Formulation

In our synthesis framework (outlined in Chapter 5), a nearly one-to-one mapping exists between control nodes in the compiler's CDFG and the synthesizable modules produced in HDL code. A high-level floorplan of these modules will give a good estimate of the on-chip distance between any given pair of modules. This means that a glance at the floorplan should determine how far or close actual basic blocks of the program are on the physical device.

Additionally, a compiler-level analysis of the high-level code can identify the actual number of variables passed between any given pair of blocks (which roughly corresponds to the amount of interconnect needed between them). The compiler decides which of these values shall be passed through centralized memory and which shall be passed via direct connection, therefore it can easily determine which signals shall be sent from any basic block to any other basic block.

Thus, from a floorplan, we should be able to determine the cost of communicating between any two basic blocks (or control nodes) in the original CDFG. If we feed this cost back to the compiler, we can re-formulate the placement of Φ-nodes as a combinatorial optimization problem designed to minimize data communication on the final synthesized circuitry.

3.2 The Φ Placement Problem

Assume that we are given a fully connected, weighted data cost graph Gcost=(Vcost, Ecost) defining the data communication cost between every basic block in the CDFG. Vcost is the exact set of CDFG control nodes, and

$$E_{cost} = \left\{ e_{ij} \mid head(e_{ij}) = v_i \wedge tail(e_{ij}) = v_j \wedge v_i, v_j \in V_{cost} \wedge i \neq j \right\}.$$

The *Φ Placement Problem* is the following. Find the set $place(\phi_a)$ (for each ϕ-function ϕ_a) such that the total communication cost C for $place(\phi_a)$ is minimized, where data cost $C(place(\phi_a))$ of a Φ-function ϕ_a is represented by

$$C(place(\phi_a)) = \sum_{p \in place(\phi_a)} \sum_{s \in S(\phi_a)} weight(e_{sp}) +$$

$$\sum_{p \in place(\phi_a)} \sum_{d \in D(\phi_a)} \begin{cases} 0 & \text{if no path exists from p to d} \\ weight(e_{dp}) & \text{otherwise} \end{cases}$$

In the above equation, each element p of set $place(\phi_a)$ is a node in the CDFG at which that Φ-function may be placed (i.e. a candidate Φ-node). The set S is the set of nodes that contain source values for the Φ-function. The set D is the set of nodes at which the name defined by the Φ-function is used. The edges e_{sp} and e_{dp} are edges of the graph G_{cost}. The solution to this problem (i.e. the sets of nodes of $place(\phi_a)$), are subject to the following constraints (Note: the notation x -+-> y is the iterated non-inclusive edge, which represents the set of edges in the directed path from x to y), which maintain the program's correctness in SSA form:

1. $\forall s \in S(\phi_a) \exists$ a control flow path s -+-> p $\forall p \in place(\phi_a)$

2. $\forall d \in D(\phi_a) \exists$ at least one $p \in place(\phi_a)$ such that p-+->d

The first constraint maintains that the definition of each source variable of a Φ-function is live at that each Φ-node defining that function. This enables proper propagation of source values to a Φ-node that it may select between them. The second constraint maintains that every destination (or use) of a Φ-function's value can be reached by at least one Φ-node (which defines that Φ-function). In other words, the variable of a Φ-function will be live at any given node that uses it. Each of these constraints is trivially required for program correctness, as we must ensure that a variable's definition may reach its use point.

Theorem 8.4: The Φ Placement Problem is NP-Complete.

Proof: Let us temporarily assume a slightly simpler version of the Φ Placement Problem, in which all edge weights in G_{cost} are unit (or 1). In this version of the problem, minimization of a placement cost C is equivalent to minimizing the number of selected nodes (that is, minimizing $| place(\phi_a)| $). Let us assume we already have a set of candidate placement nodes which meet the first constraint of the Φ Placement problem. Finding these nodes on a flow graph is a simple task, as it involves finding the nodes which are reachable from every source of the Φ-function. (To model the spatial placement option, we simply include each destination node as a candidate node.) The problem then involves finding a minimum set cover of the destination nodes in D using the candidate placement nodes that can reach them. In other words, the algorithm selects the minimum set of candidate placement nodes such that each destination node in D can be reached by one or more nodes in the chosen placement set $place(\phi_a)$. Since this instance of the Φ Placement problem must be as hard as minimum set cover, and minimum set cover is NP-Complete, the Φ Placement problem is NP-Complete. □

Corollary 8.4.1.: The Φ Placement problem is NP-Complete even when all inter-node data communication is equivalent (or unit) cost.

Proof: Follows immediately from proof of Theorem 6.4. □

3.3   Solution to the Φ Placement Problem

The solution to this problem requires finding and measuring the cost of various possibilities of $place(\phi)$ for a given Φ-function. These possibilities must adhere to the correctness constraints mentioned in the previous section. The permuted list of possible Φ placements has been proven to be exponential (as Φ-nodes may be moved and/or replicated from the original iterated dominance frontier position, and these replicas themselves may be moved and/or replicated, etc.). However, some intuitive observations about the nature of the placement on a two-dimensional chip leads to some potentially useful pruning of the search space:

Theorem 8.5: If a Φ-function's destination variable is only used in one control flow node (i.e. if $|D| = 1$), then the Φ-function should not be replicated such that there exist as many nodes of this Φ-function as there are sources of the Φ-function (i.e. $|place(\phi)| < S$).

Proof: This can be proven via triangle inequality in a planar two-dimensional placement simulating our final chip and the positions of Φ-nodes, sources, and destinations thereon. We can model the planar two-dimensional placement of our scenario ($|D| = 1$, $|S| = |place(\phi)|$) as a set of triangles, where an edge is drawn from each source to its closest Φ, from each Φ to the destination node, and from each source to the destination node. In Figure 63, this is demonstrated by a scenario in which there is one destination, two source nodes, and the Φ-node is replicated twice. In the figure, additional non-solid edges of communication are drawn between each source and the Φ-nodes which are not closest to it. Triangle inequality states that sum of the lengths of two sides of any triangle is longer than the third side. By triangle inequality, the sum of the cost of the paths from the sources to the destination will be shorter than the paths through the Φ-nodes, which must contain at least each source->Φ-node->dest path. Additionally, each source must communicate to every other Φ-node via the non-solid edges. Obviously, it is much cheaper (in terms of wirelength) to communicate from each source to the destination (which would only use a single Φ-node, placed spatially). This planar triangle formulation can be built for any number of $|S| = |place(\phi)|$ and $|D| = 1$, no matter where the sources, destinations, and nodes are placed. In all cases, the option exists to use fewer Φ-nodes than sources (by placing a Φ-node spatially at the single destination). $\square$

Due to the necessity of building communication wires from all sources to all Φ-nodes (and from these Φ-nodes to a set of destinations), replication of Φ-functions (i.e. placement at many Φ-nodes) may hurt more than help in the general case. Additionally, the range of possible placements is not very large for a given Φ-function, as Φ live ranges tend to be short, and Φ functions tend not to have many sources. When considering placement options, the number of times a Φ may be duplicated in a placement should be constrained by a constant k. This constant will practically limit the number of Φ-nodes allowed in a given placement *place*(ϕ). By constraining $|place(\phi)| < k$ (for a chosen k), we do limit our design space, but we are also placing a pragmatic limitation on the amount of data communication required. For this work, k was chosen to be 3.

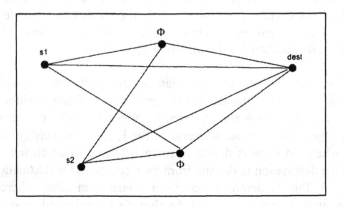

Figure 63 An example of Theorem 6.5: Triangle Inequality in placement

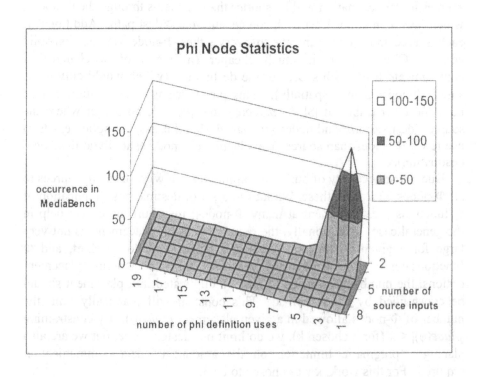

Figure 64 Number of sources, dests for Φ nodes in our benchmark suite

This intuition is empirically backed by experiments shown in Figure 64. This figure represents the number of sources and number of destinations of Φ-nodes in our benchmark suite (by frequency of occurrence). As demonstrated, most Φ-nodes have only two sources, and almost all of them have less than four destinations. For these cases, which represent the majority, it would be wasteful to consider Φ placements with many duplicates of the Φ function. Thus, we feel that our restriction of |*place*(φ)| is economical in terms of number of wires and multiplexers used.

3.4   Updated Φ Placement Algorithm

The new Φ Placement Algorithm in Figure 65 builds a set of placement options for each Φ-function. These placement options are limited to k CDFG nodes per placement. For instance, for a Φ-function whose placement options are { {1, 2}, {5}, {3,6,8} }, the Φ-function can either be placed at CFG node 5, replicated and placed at CFG nodes 1 and 2, or replicated and placed at CFG nodes 3, 6, and 8. In this case, if k=2, then the {3,6,8} option would not be considered, and would be dropped from the list of placement options. As stated in lines 10-11 of the algorithm in Figure 65, the placement options are found using a search function, which starts execution at the iterated dominance frontier of the Φ-function's sources. After the set of possible placement options are discovered by the function findPlacementOptions, each of the placement options is evaluated via the cost function described in Section 3.2. The best Φ placement option is then used, and the Φ-node is distributed and duplicated to this final placement. The algorithm's internal representation of each placement option is a set of numbers, where each number represents a CDFG node where the Φ-function must be placed.

1. Given a CDFG $G(N_{cfg}, E_{cfg})$
2. perform_SSA(G)
3. calculate_def_use_chains(G)
4. remove_back_edges(G)
5. topological_sort(G)
6. **for each** node $n \in N_{cfg}$
7.   **for each** Φ-node $\Phi \in n$
8. $s \leftarrow \Phi.sources$
9. $d \leftarrow |def\_use\_chain(\Phi.dest)|$
10. $IDF \leftarrow$ .iterated dominance frontier of s
11.   *PossiblePlacements* $\leftarrow$ findPlacementOptions(*IDF*)
12.   place(Φ)$\leftarrow$ selectBest (*PossiblePlacements*)
13.   distribute/duplicate Φ to place(Φ)

Figure 65 Updated Φ Placement Algorithm

The function findPlacementOptions is a recursive algorithm which generates the possible placements for each Φ-function. To generate the set of sets which lists the placement options, findPlacementOptions uses dominance information to traverse the graph in top-down fashion. (It is assumed for this algorithm's purposes that there are no unnatural loops in the code. A modification to the algorithm would need to be made otherwise.) findPlacementOptions is shown in Figure 67.

On line 10 in Figure 67, the findPlacementOptions algorithm makes use of a binary function named crossProductJoin, shown in Figure 68. crossProductJoin takes two sets of sets and combines them, removing duplicates and sets which are supersets of other sets. This removal of supersets ensures that the Φ-function is only placed at as many nodes as necessary. For example, crossProductJoin({{3}, {4,5}}, {{2}, {6}}) returns {{2,3},{2,4,5},{3, 6},{4,5,6}}. However, crossProductJoin({{2}, {3}}, {{3}, {4,5}}) produces {{2,3}, {2,4,5}, {3}, {3,4,5}} before line 8 of the algorithm. After sets which are supersets of other sets are removed ({2,3} and {3,4,5} are supersets of {3}), the function finally returns {{2,4,5}, {3}}. Note that by dropping {2,3} and {3,4,5} from consideration, we remove unnecessary placement options. Put differently, if the existance of {3} means we have the option to place a Φ at CDFG node 3, then the existance of a superset means we also have the option to place a Φ node in some other places and also at node 3. There is no need to place the Φ-node redundantly (as this wastes multiplexers and wirelength), so we disregard all sets which are supersets of any other sets.

In addition to the aforementioned functionality, crossProductJoin immediately drops any sets from the generated set of sets which are larger than k (i.e. would correspond to a placement of > k Φ-nodes).

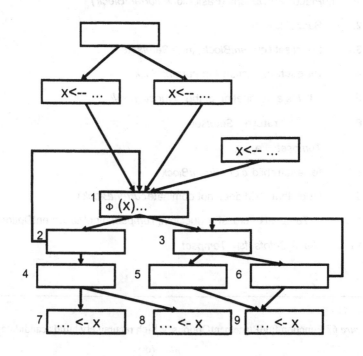

Figure 66 Example Portion of CDFG

To clarify the functionality of findPlacementOptions, we present a small example of the function in action on a portion of a CDFG. The portion of the CDFG pertinent to this example is shown in Figure 66. In this example, k = 3. Execution would proceed as follows:

findPlacementOptions(7) returns { {7} }
findPlacementOptions(8) returns { {8} }
findPlacementOptions(4) returns { {7, 8}, {4} }
findPlacementOptions(2) returns { {7, 8}, {4}, {2} }
findPlacementOptions(9) returns { {9} }
findPlacementOptions(6) returns { {6}, {9} }
findPlacementOptions(5) returns { {5}, {9} }
findPlacementOptions(3) returns { {6, 5}, {9}, {3} }
(Note that {6, 9} and {9, 5} are supersets of {9} and will be removed.)

findPlacementOptions(1) returns { {6, 5, 4}, {6, 5, 2}, {7, 8, 9}, {4, 9}, {2, 9}, {3, 7, 8}, {3, 4}, {2, 3}, {1} }

---

1. findPlacementOptions (BasicBlock *currentBlock*)

2.   *Setofsets*← ø

3.   Insert set {*currentBlock*} into *Setofsets*

4.   **for each** instruction *I* in *currentBlock*

5.   **if** *I* is a destination of the Φ function   *Φ*

6.   **return**   *Setofsets*

7.   *Tempset*← ø

8.   **for each** child *c* of *currentBlock*

9.   (such that child does not dominate *currentBlock*)

10.   *Tempset*← crossProductJoin(*Tempset* findPlacementOptions(*c*))

11.   **return** *Setofsets*∩ *Tempset*

Figure 67 Function findPlacementOptions, which recursively builds candidate sets for

place(Φ)

(Note that cross product element {5, 6, 7, 8} was disallowed because its cardinality is greater than $k = 3$.)

The results of findPlacementOptions(1) is the final set of Φ placement options for the entire Φ-function (as CDFG node 1 is the iterated dominance frontier of the set of sources for variable *x*.)

3.5   Floorplan Cost Model

As mentioned in Section 3.2, the Φ Placement Problem assumes an input cost graph $G_{cost}=(V_{cost}, E_{cost})$ where each edge $e \in E_{cost}$ is weighted with the cost of communication between the two vertices that it connects. The graph can also be represented (and was actually implemented) as an nxn cost matrix, where n is the number of control nodes (basic blocks) in the CDFG and Cost[i, j] is the cost of communicating between CDFG nodes i and j. Note, in general, that Cost[i, j] = Cost[j, i]. Also, Cost[i, i] = 0, as we do not model the local cost of communication within a basic block in a CDFG, and therefore consider it to be cost-free.

```
1.  crossProductJoin ( setofsets set1, setofsets set2) {
2.      if set1 is empty, return set2
3.      if set2 is empty, return set1
4.      Newset ← ø
5.      for each set s in set1
6.         for each set t in set2
7.            Newset ← Newset ∪ { s ∪ t}
8.      Remove all sets from Newset that are
9.            supersets or duplicates of other sets in Newset
10.     return Newset
```

Figure 68 Function crossProductJoin, a binary operation to join two sets of sets

The cost between a given pair of CDFG nodes has two components: the distance between the nodes on the floorplan, and the amount of data communicated between these two nodes using wires. The distance between a pair of nodes on the floorplan is taken using the Euclidean (center-to-center) distance between the modules. This can easily be obtained using simple geometry on the floorplanner's output file (which specifiesthe size and coordinates of each module in the floorplan). The amount of data communicated between two nodes is the sum of the total bits of communication passed between these nodes. For instance, if CDFG nodes share variables i and j (each 32-bit integers), and one ASCII character c (8 bits), then the amount of communication passed between these nodes is 72 bits.

The formula representing the communication cost between any two basic blocks i and j (with Euclidean distance *Euclidean(i,j)* and total communication amount *bits(i,j)*) is as follows:

$$Cost[i,j] = \begin{cases} Euclidean(i,j) * bits(i,j) & if\ bits(i,j) > 0 \\ Euclidean(i,j) * 32 & otherwise \end{cases}$$

The second case, which substitutes number of bits communicated for the constant 32, derives from the following anomaly: how can we determine the communication cost between two CDFG nodes if they are not yet sharing any data? In this case, bits(i,j) = 0. However, sharing data between these CDFG nodes still incurs a design cost. This cost must still depend on the distance between the two CDFG nodes on the final floorplan. For this cost to appear significant in the face of the other costs, a multiplier is needed which will make this cost commensurate with costs between nodes which share signals. The constant 32 was chosen because it is the most often exhibited size of bits shared between any two CDFG nodes (i.e. most nodes share data the size of a single integer variable).

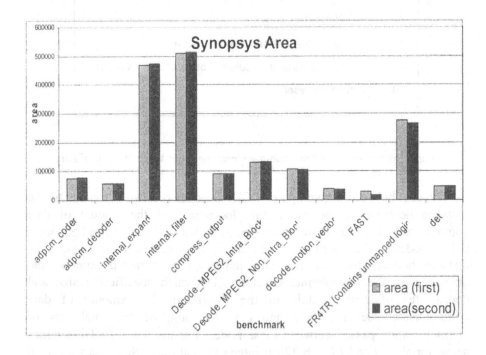

Figure 69 Comparison of the total area of the first and second synthesis iterations as measured by Synopsys

4.   FEEDBACK EXPERIMENTAL RESULTS

To measure the effectiveness of introducing feedback into the design flow, we compiled the same DSP functions from MediaBench into behavioral and structural VHDL, and then synthesized the designs in the Synopsys Behavioral and Design Compilers (as described in Section 2). We then took the individual areas of each design module (corresponding to control nodes from the CDFG) from Synopsys, and extracted a data file listing the modules and their possible heights and widths. For each module, we took the total area reported by Synopsys and used this to calculate and insert five aspect ratio options (1:1 (square), 2:1, 4:1, 1:2, and 1:4) into the data file. We then created a net file using information from our behavioral VHDL conversion pass. The net file listed the connections and connection-widths (in bits) between CDFG nodes. These files were supplied to our floorplanner [296], which was then executed to create a working floorplan. The output of the floorplanner was converted into a cost matrix as described in Section 3.5. This concluded the *first synthesis iteration*.

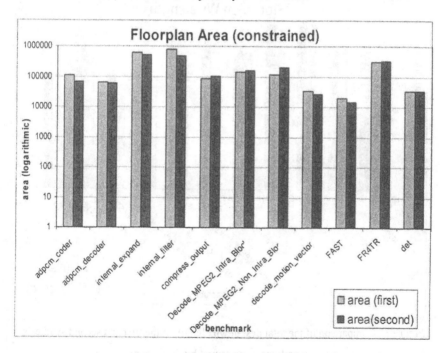

Figure 70 Comparison of the total design area of the first and second synthesis iterations as measured by floorplanner

Next the compiler pass performing SSA was re-executed upon each benchmark, this time with the each benchmark's cost matrix from the first synthesis iteration. Φ-nodes were replicated and distributed as described in Sections 3.3 and 3.4. Then the conversion to VHDL and subsequent behavioral/logic synthesis in Synopsys was repeated with the new Φ placement. New data and net files were created from this synthesis, and the floorplanner was re-executed. (This time the floorplanner was restricted to use the same aspect ratio for each module that it had selected in the previous iteration.) This creation of the final floorplan concluded the *second synthesis iteration*

Figure 69 compares the total area estimate provided by Synopsys Design Compiler during the first and second iterations of synthesis. Our Φ placement provides no clear benefit to overall estimated area, as the results are mixed. Area tends to stay nearly the same between design iterations, with the notable exception being FR4TR. However, this design was not fully completed by Synopsys, and its results can be considered insignificant.

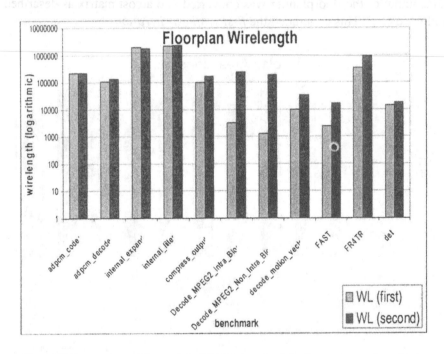

Figure 71 Comparison of the total design wirelength of the first and second synthesis
iterations as measured by floorplanner

Figure 70 and Figure 71 represent the floorplanner's estimate of area and wirelength of the design iterations (respectively). While a small but measurable area decrease occurs for every benchmark (most notably in the larger benchmarks internal_expand, internal_filter, and FR4TR), wirelength is strictly non-decreasing. In fact, in every case but one (adpcm_coder), wirelength increases. Sometimes the increase in wirelength is quite drastic, as   in   the   cases   of   Decode_MPEG2_Intra_Block,

Decode_MPEG2_Non_Intra_Block, and FAST. These results are anomalous, as we intuit that the redistribution of Φ-nodes will reduce overall interconnect in the design.

Table 6 Aspect Ratios of floorplan during first and second iterations of synthesis

Benchmark	AR (first)	AR (second)	Delta Aspect Ratio
adpcm_coder	1.271	1.081	0.191
adpcm_decoder	1.424	1.072	0.351
internal_expand	1.597	1.137	0.461
internal_filter	1.063	1.346	0.284
compress_output	1.432	1.102	0.330
Decode_MPEG2_Intra_Block	1.096	1.329	0.234
Decode_MPEG2_Non_Intra_Block	3.369	1.112	2.257
decode_motion_vector	2.127	1.417	0.710
FAST	2.326	1.831	0.495
FR4TR	2.041	1.002	1.039
det	2.517	1.327	1.190

In order to understand the seeming failure of the design feedback loop, it is important to understand exactly how much information the feedback loop has given us. Although the compiler has made an informed Φ placement decision with knowledge about the design's physical characteristics, the decisions are based upon the characteristics of the first synthesis iteration. If the second synthesis iteration is sufficiently different than the first iteration, then these decisions may not lead to data communication reduction. In other words, a large difference between the first and second synthesized designs may lead to unexpected results in feedback-driven optimization. Certainly a small change in some module areas (via the removal/addition of multiplexers for Φ-nodes) can result in an entirely different floorplan for the second design iteration. To see how drastic this difference can be, refer to Table 6, which demonstrates the difference in aspect ratios between the first and second floorplans. Not surprisingly, the most pronounced floorplan difference (Decode_MPEG2_Non_Intra_Block) also corresponds to the largest increase in wirelength in Figure 71.

Data Communication

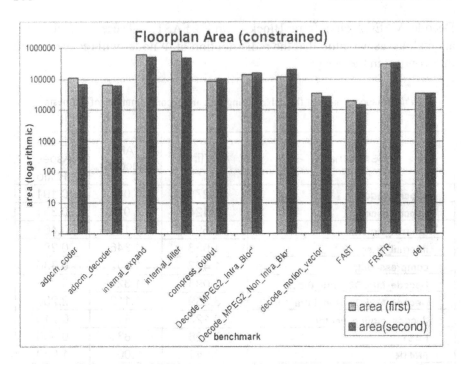

Figure 72 Comparison of the total design area of the first and second synthesis iterations as measured by floorplanner with an aspect ratio constraint

Table 7 Aspect Ratios of floorplan with aspect ratio constraint

benchmark	AR (first)	AR (second)	Delta aspect ratio
adpcm_coder	1.27	1.27	0.003
adpcm_decoder	1.42	1.23	0.194
internal_expand	1.60	1.52	0.074
internal_filter	1.06	1.03	0.028
compress_output	1.43	1.01	0.423
Decode_MPEG2_Intra_Block	1.10	1.10	0.003
Decode_MPEG2_Non_Intra_Block	3.37	3.37	0.001
decode_motion_vector	2.13	1.67	0.460
FAST	2.33	2	0.326
FR4TR	2.04	1.78	0.258
det	2.52	2.13	0.391

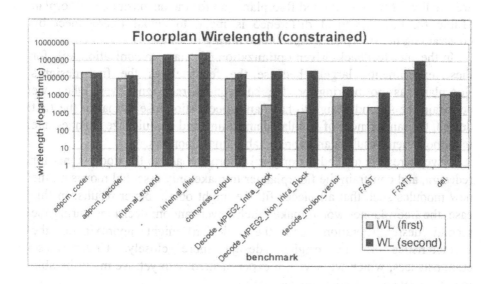

Figure 73 Comparison of the total design wirelength of the first and second synthesis

iterations as measured by floorplanner with an aspect ratio constraint

In an attempt to make the second floorplan very similar to the first floorplan, and thus maintain our circuit characteristics across redesign, we repeated the original set of experiments, this time constraining the floorplanner. Specifically, we used a switch to request that the floorplanner attempt to produce an overall aspect ratio as close to that of the original design as possible. This constrained floorplan creation in the second design iteration resulted in the area, wirelength, and aspect ratio measurements of Figure 72, Figure 73, and Table 7 respectively. The overall design area measured by each iteration of the floorplanner changes only slightly between iterations, a much more subtle difference than in the original experiments of Figure 70. However, when the floorplanner is constrained, the area sometimes increases between iterations (most notably in Decode_MPEG2_Non_Intra_Block) whereas the unconstrained floorplanner always produced an area reduction between iterations. The results of design wirelength between iterations in Figure 73 is also disappointing, as the wirelength increased in all cases but one (adpcm_coder). A quick comparison of Figure 73 and Figure 71 demonstrates that the wirelength results are nearly identical whether the floorplanner is constrained or not.

Even though the differences in aspect ratio between first and second designs are smaller for the constrained floorplan than for the unconstrained floorplan (Table 6), little practical difference is made to actual interconnect by constraining the floorplan's aspect ratio during redesign.

In the end, feedback-driven optimization of data communication fails for these experiments, largely because the feedback of the first design's floorplan has no bearing on the subsequent redesign. Even when we constrain the second design to have an aspect ratio close to that of the first design, the randomness of simulated annealing and module size changes due to Φ movement result in an unpredictable final floorplan.

If we were able to send the original floorplan back to the floorplanner for redesign, and constrain the floorplanner to make only essential moves on the new modules such that all pieces fit, we might obtain better results. In this case, the floorplanner would make much fewer random decisions during the second design iteration, and the redesign might approximate the characteristics of the original design more closely. Under these circumstances, which we have not experimented with yet, we might be able to reduce data communication successfully.

However, many questions exist regarding this approach. If we constrain the floorplanner, we restrict it from performing its own optimizations. How much optimality do we lose by restricting the floorplanner compared to the potential reduction in data communication that we gain? Also, every time we redistribute Φ-nodes, we are changing module sizes of affected basic blocks. Is it always possible to make the second floorplan very close to the original design? If not, how close is "close enough?" Obviously, further research is needed to explore these gray areas.

5.   RELATED WORK

Many compiler techniques use SSA for analysis or transformation [297-299]. Also, there have been modifications of SSA form [174, 300]. To the best of our knowledge, this was the first work to consider SSA form for hardware compilation. Several recent reconfigurable system compilers (e.g.[206, 207]) use the notion of SSA, though they do not provide any analysis of the effect of SSA on the final circuit. SA-C [209] proposes a single assignment language by definition. Although this restriction on the SA-C language may make it unwieldy for many human programmers, we have seen that high-level code can be automatically transformed into single assignment form. It may even be possible to use our SSA techniques as a front end to this language.

6. SUMMARY

In this work, we presented methods needed for hardware compilation. First, we described a framework for compiling a high-level application to an HDL. The framework includes methods for transforming a traditional compiler IR to an RTL-level HDL. We illustrated how to transform the IR into a CDFG form. Using the CDFG form, we explained methods to control the path of execution. Furthermore, we gave methods for communicating data between the control nodes of the CDFG.

We examined the use of SSA to minimize the amount of data communication between control nodes. We showed a shortcoming of SSA when it is applied to minimizing data communication. The temporal positioning of the Φ-node is not optimal in terms of data communication. We formulated an algorithm to spatially distribute the Φ-node to minimize the amount of data communication. We showed that this spatial distribution can decrease the data communication by 20% for some DSP functions. Additionally, we proved that if all data communication wire-lengths are of unit cost, the Spatial SSA Algorithm provides minimal data communication.

However, actual area measurements taken from a synthesis tool indicated that our TEW model of communication was oversimplified. We reformulated the communication cost model to more clearly express the actual wirelength of inter-control node communication on the final circuit. We also reformulated the Φ Placement Problem as a data communication optimization problem: minimize the total data communication of the program subject to code correctness constraints. Finally, we redesigned the Φ Placement algorithm to accept a cost matrix as input. In other words, we redesigned our compiler optimization to accept information from a later stage of synthesis, introducing a feedback loop to our design flow.

Experiments with this new design flow were erratic, and tended not to demonstrate any reduction in wirelength (even though for many experiments, circuit area decreased). These unpredictable results occurred because the first synthesized circuit (that upon which cost estimates were made) was very different from the second synthesized circuit. This was the case even when the floorplanner was constrained to the original aspect ratio on its second execution. Clearly more research is needed to accommodate design feedback loops with constrained floorplans, where the second floorplan is made as close to the original as possible. This will introduce a new tradeoff to the design space: a potential sacrifice of objectives at the floorplanner (which will be constrained during the second design iteration) in order to aide data communication optimization at the compiler level. Such a discussion is beyond the scope of this book.

Chapter 9

INCREASING HARDWARE PARALLELISM

1. INTRODUCTION

Hardware has a distinct advantage over software as operations can be spatially computed; this allows parallelism among the operations. Unfortunately, an application programmer does not always exploit all of the parallelism available in the application. We must employ techniques that discover additional parallelism so that the hardware performance is maximized.

The number of operations in a basic block – hence an initial control node – can be quite small. Studies show that the average instruction level parallelism (ILP) per basic block is between 2 – 3.5 [301]. By combining basic blocks, one can increase the amount of ILP. There are many compiler optimizations that combine basic blocks such as trace scheduling [14], superblock [172] and hyperblock formation [187].

2. TRACE SCHEDULING AND SUPERBLOCKS

Compilers for superscalar and VLIW processors increase parallelism by combining frequently executing sequences of basic blocks. Combining basic blocks allows advanced scheduling and optimization by ignoring the control constraints associated with the alternate paths of execution. *Trace scheduling* finds a *complete trace* – a path from the start to the end of the CDFG – and combines the basic blocks on that path.

Trace scheduling incurs a large amount of complexity (particularly during scheduling) in order to maintain correct program execution. In particular, side entrances to the trace create an immense amount of bookkeeping. The superblock eliminates side entrances; this makes scheduling and other compiler optimizations much easier. A *superblock* is a set of basic blocks from one path of control where execution may only begin at the top but may leave at one or more exit points.

Every method that we discussed to increase the parallelism combines control nodes so that the operations in these nodes can be executed in parallel. There are many proposed methods for combining control node sequences. A runtime method profiles branching characteristics of the application. Based on this information, commonly occurring control node sequences are combined. This requires input sequences, which may not always be available. Also, different input sequences may cause drastically different profile information. But, when applications that have commonly occurring, well-represented input sequences, the profile information is immensely helpful for combining control nodes.

Hank et al. [302] develop a static method for superblock formation. They use heuristics to predict the commonly taken branches through the program structure. Using this branch frequency prediction, they combine basic blocks without hazards. A hazard is an instruction or group of instructions whose side effects may not be completely determined at compile time e.g. I/O instructions, subroutine calls, jumps with indirect target addresses. They show that static methods can rival the dynamic techniques.

There are many factors to consider during superblock formation. First and foremost, we wish to increase the parallelism. By increasing the number of operations in the superblock, we increase the opportunity for parallelism. But, there are several factors that may reduce the amount of parallelism we can achieve. For instance, a large amount of dependencies between operations in different control nodes of the superblock will limit the number of instructions that we can schedule at a particular time. Therefore, we want to minimize the length of the critical path of the superblock.

When we consider a hardware implementation of the application, the selection of control nodes to include in the superblock can have a direct effect on minimizing the amount data communication. By combining basic blocks into superblocks or traces, we are effectively choosing a hardware partitioning. The control nodes in the superblock must remain in the same vicinity or partition. Additionally, the data communication between the nodes of a superblock becomes local. *Local communication* – defined as data communication between control nodes in the same partition – has smaller delay than *global communication* – the data communication between partitions.

3.   HARDWARE PARTITIONING USING TRACE SCHEDULING AND SUPERBLOCKS

In this section, we describe two problems associated with increasing the parallelism of the hardware. We wish to create a trace schedule or set of superblocks that produce a good partitioning of the application. In particular, we attempt to maximize the local interconnect of the hardware. This is accomplished by partitioning the operations such that the data communication within the partitions is maximized. In essence, we attempt to maximize the amount of local communication by placing control nodes that have a lot of communication between them in the same superblocks or a trace.

Given a CDFG $C(N, E_c, E_d)$, each node $n \in N$ has a weight w_n. The vertex weight w_n represents the amount of area occupied by an implementation of the computations in the node. We can use actual synthesized node as the area or some type of area estimation. The CDFG has two distinct sets of directed edges, E_c corresponds to the control flow constraints and E_d corresponds to the data communication between the nodes. There is a weight w_e for each $e_d \in E_d$. The edge weight w_e of edge $e(n_i, n_j)$ (i and j are two different control nodes) corresponds the amount of data that is being transferred along that edge (data being transferred from control node i to control node j). We wish to partition the CDFG such that the sum of the node weights within a partition is less than some constant. Additionally, the nodes in a partition must be superblock or complete trace. Furthermore, we attempt to maximize the sum of the edge weights between any two nodes in the same partition.

By partitioning the control nodes, we accomplish two things. First, the edges that are local to a partition have local data communication. Local communication has smaller delay than global communication. By maximizing the edge weights within the partitions, we minimize the amount of global delay. Second, by restricting the partition to a set of control nodes that form a superblock or complete trace, we increase the parallelism in the CDFG. The area restriction on the partition insures that the CDFG for each partition has approximately the same size. We can use this to limit the amount of operations that can be optimized by data flow synthesis techniques. Also, this allows us to keep the silicon area of the each control node under a specified constraint.

In addition to area restrictions on the partitions, we could have parallelism limits. Ideally, control nodes would have a large amount of parallelism so that the data flow scheduler would have the freedom use take advantage of that parallelism, if it so desires. In order to realize this, we must have a method to determine the amount of parallelism present in a control node. A quick metric could use an ASAP scheduling. Then, we could determine the amount of parallelism at each time step. From that, an "average" parallelism can be determined. This method gives an upper bound on the average amount of parallelism. It also gives a lower bound on the delay. A more complex, yet more accurate metric could use one of the many data flow scheduling algorithms. Since trace scheduling must form a complete trace, it may not always allow an area constraint.

The trace scheduling partitioning and superblock partitioning problems can be formulated as follows:

Trace Scheduling Partitioning Problem Definition: Given a CDFG and a minimum parallelism constraint P, determine a complete trace that maximizes the data communication between the nodes of the trace such that the parallelism of the trace is greater than P.

Superblock Partitioning Problem Definition: Given a CDFG, a minimum parallelism constraint P and a maximum area constraint A, determine a set of superblocks that maximizes the local communication such that the parallelism of each superblock is greater than P and the combined node area of superblock is less than A.

In the next section, we show the effectiveness of trace scheduling and superblock formation for hardware partitioning.

4.   EXPERIMENTAL RESULTS

We developed two greedy algorithms for trace scheduling and superblock formation. The algorithms greedily select control nodes based on data communication. The trace scheduling algorithm starts at the entry control node and iteratively chooses the next control node of the trace based on the successor control node that has the greatest amount communication data with the nodes that are already in the trace. The superblock formation algorithm iteratively creates superblocks. An initial node for the superblock is selected by finding the node (which is not already in another superblock) with the greatest amount of data communication to other nodes that are not currently a part of another superblock. Then, the algorithm proceeds like the trace scheduling algorithm until it's successors are all part of another superblock or we reach the area limit. We use a function of additional area added by a node to the superblock and the amount of additional data communication added to the superblock in order to choose the next node to add to the superblock.

We ran the two algorithms on the benchmarks described in Table 3.1 in Section 3.6. We look at the number of operations per partition and the amount of local data communication after partitioning for the two algorithms. As in the previous section, we use the data communication edge weight as a measure of the data communication. First, we look at the number operations in each partition. The number of operations relates to the area of the partition. Since a trace schedule must form a complete trace, it is much harder to control the number of operations. When we use superblocks for partitioning, we can easily add an area constraint to the problem formulation since the superblock does not have the complete trace restriction.

Figure 74 displays the number of operations per partition for trace scheduling and superblock formation. For trace scheduling, the percentage is simply the number of operations in the trace compared to the total number of operations. For superblocks, the percentage is the average number of operations per superblock compared to the total number of operations. The number of operations in the trace can be quite large; the trace consumes approximately 80% in the adpcm1 benchmark. This mainly stems from the fact that there is no effective way of limiting the operations because we must have a complete trace. In our experiments, we choose limit the area of the operations for superblock scheduling. Most of the benchmarks had an average area under 20%. The two exceptions are fft1 and motion. Both of these benchmarks had a few nodes with a large amount of operations in them. The greedy algorithm chose to include these nodes as superblocks because they had a large amount of data communication going to and from them.

The main objective of partitioning is to maximize the amount of local communication. Figure 75 shows the amount of local communication for both trace scheduling and superblock formation. Over all the benchmarks, trace scheduling converted an average of 12.7% of the total communication to local communication. On average, superblock formation localized 17.5% of the overall data communication. For the benchmarks where trace scheduling has more local communication than superblocks, the number of operations in the trace is inordinately large. For example, the trace scheduling partitioning of adpcm1, getblk1, getblk2, and fft1 outperform the superblock partitioning. This is mainly because the traces have a large amount of the total operations in them. Referring to Figure 74 each of these traces has over 50% of the total operations in it. This is unacceptable as a partition. Even though we restrict the number of operations in the superblocks, we can still get a good partitioning in terms of maximizing local data communication.

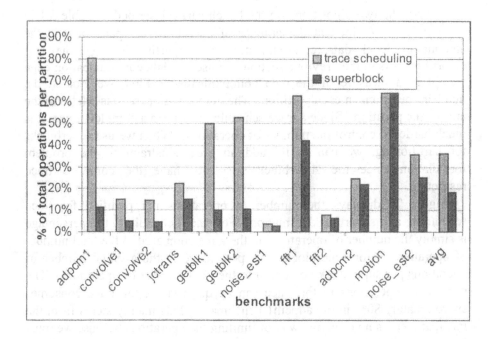

Figure 74 The percentage of operations per partition for trace scheduling and superblock formation.

Our results show that superblock formation is much better technique for increasing parallelism during hardware partitioning. Additionally, the side entrances of trace scheduling constrain the scheduler. Superblocks eliminate side entrances; hence they do not have such a problem. Therefore, we believe that superblocks are a far more effective structure for increasing parallelism during hardware partitioning.

5.   RELATED WORK

The idea of hardware compilation has been discussed since the 80's. At that time, it was under the guise of silicon compilation and related closely to what is referred to as behavioral synthesis nowadays.

The past 15 years have brought about a number of platforms that take high-level code and generate a hardware configuration for that platform. The PRISM project [303] took functions implemented in a subset of C and compiled them to their FPGA-like architecture. The Garp compiler [29] takes automatically maps C code to their MIPS + FPGA architecture. The DeepC compiler [213] is the most similar to our work, as it creates

results show that the superblock is much more effective than trace scheduling for this purpose.

Acknowledgments

This book would have been impossible without our collaboration with number of people. In particular, we would like to thank Kia Bazargan, Elaheh Bozorgzadeh, Philip Brisk, Wenrui Gong, Seda Ogrenci Memik, Ankur Srivastava and Gang Wang.

References

[1] M. J. Wirthlin and B. L. Hutchings, "A dynamic instruction set computer," *Proceedings of the IEEE Symposium on FPGAs for Custom Computing Machines*, 1995.

[2] S. Hauck, "The roles of FPGAs in reprogrammable systems," *Proceedings of the IEEE*, vol. 86, pp. 615-38, 1998.

[3] D. A. Buell and K. L. Pocek, "Custom computing machines: an introduction," *Journal of Supercomputing*, vol. 9, pp. 219-29, 1995.

[4] A. DeHon, "Comparing computing machines," *Proceedings of the Configurable Computing: Technology and Applications*, 1998.

[5] J. E. Vuillemin, P. Bertin, D. Roncin, M. Shand, H. H. Touati, and P. Boucard, "Programmable active memories: reconfigurable systems come of age," *IEEE Transactions on Very Large Scale Integration (VLSI) Systems*, vol. 4, pp. 56-69, 1996.

[6] A. DeHon and J. Wawrzynek, "Reconfigurable computing: what, why, and implications for design automation," *Proceedings of the Design Automation Conference*, 1999.

[7] E. Mirsky and A. DeHon, "MATRIX: a reconfigurable computing architecture with configurable instruction distribution and deployable resources," *Proceedings of the IEEE Symposium on FPGAs for Custom Computing Machines*, 1996.

[8] D. Gajski and R. Kuhn, "Guest Editors' Introduction: New VLSI Tools," *IEEE Computer*, vol. 16, pp. 11-14, 1983.

[9] G. Moore, "Cramming more components onto integrated circuits," *Electronics*, vol. 38, 1965.

[10] K. Keutzer, A. R. Newton, J. M. Rabaey, and A. Sangiovanni-Vincentelli, "System-level design: orthogonalization of concerns and platform-based design," *IEEE Transactions on Computer-Aided Design of Integrated Circuits and Systems*, vol. 19, pp. 1523-43, 2000.

[11] M. R. Garey and D. S. Johnson, "Computers and intractability. A guide to the theory of NP-completeness," 1979.

[12] T. C. Hu, "Parallel Sequencing and Assembly Line Problems," *Operations Research*, pp. 841-848, 1961.

[13] P. G. Paulin and J. P. Knight, "Force-directed scheduling for the behavioral synthesis of ASICs," *IEEE Transactions on Computer-Aided Design of Integrated Circuits and Systems*, vol. 8, pp. 661-79, 1989.

[14] J. A. Fisher, "Trace scheduling: a technique for global microcode compaction," *IEEE Transactions on Computers*, vol. C30, pp. 478-90, 1981.

[15] R. Potasman, J. Lis, A. Nicolau, and D. Gajski, "Percolation based synthesis," *Proceedings of the ACM/IEEE Design Automation Conference*, 1990.

[16] G. De Micheli, *Synthesis and optimization of digital circuits*. New York: McGraw-Hill, 1994.

[17] F. J. Kurdahi and A. C. Parker, "REAL: a program for REgister ALlocation," *Proceedings of the ACM/IEEE Design Automation Conference*, 1987.

[18] M. Sarrafzadeh and C. K. Wong, *An introduction to VLSI physical design*. New York: McGraw Hill, 1996.

[19] N. A. Sherwani, *Algorithms for VLSI physical design automation*, 3rd ed. Boston: Kluwer Academic Publishers, 1999.

[20] O. T. Albaharna, P. Y. K. Cheung, and T. J. Clarke, "On the viability of FPGA-based integrated coprocessors," *Proceedings. IEEE Symposium on FPGAs for Custom Computing Machines (Cat. No.96TB100063). IEEE Comput. Soc. Press. 1996*, pp. 206-15.

[21] A. DeHon and J. Wawrzynek, "Reconfigurable computing: what, why, and implications for design automation," *Proceedings 1999 Design Automation Conference (Cat. No. 99CH36361). IEEE. 1999*, pp. 610-15.

[22] A. DeHon, "Reconfigurable Architectures for General-Purpose Computing," in *Department of EECS*: MIT, 1996.

[23] K. Compton and S. Hauck, "Reconfigurable computing: a survey of systems and software," *ACM Computing Surveys*, vol. 34, pp. 171-210, 2002.

[24] M. Gokhale, W. Holmes, A. Kopser, S. Lucas, R. Minnich, D. Sweely, and D. Lopresti, "Building and using a highly parallel programmable logic array," *Computer*, vol. 24, pp. 81-9, 1991.

[25] P. M. Athanas and H. F. Silverman, "Processor reconfiguration through instruction-set metamorphosis," *Computer*, vol. 26, pp. 11-18, 1993.

[26] J. R. Hauser and J. Wawrzynek, "Garp: a MIPS processor with a reconfigurable coprocessor," *Proceedings of the 5th Annual IEEE Symposium on Field-Programmable Custom Computing Machines*, 1997.

[27] R. Razdan and M. D. Smith, "A high-performance microarchitecture with hardware-programmable functional units," *Proceedings of the 27th Annual International Symposium on Microarchitecture. MICRO 27. ACM. 1994*, pp. 172-80.

[28] S. Hauck, T. W. Fry, M. M. Hosler, and J. P. Kao, "The Chimaera reconfigurable functional unit," *Proceedings of the Symposium on Field-Programmable Custom Computing Machines*, 1997.

[29] T. J. Callahan, J. R. Hauser, and J. Wawrzynek, "The Garp architecture and C compiler," *Computer*, vol. 33, pp. 62-69, 2000.

[30] H. T. Kung and C. E. Leiserson, "Systolic arrays for VLSI," *Sparse Matrix Proceedings. Philadelphia, PA: Society of Industrial and Applied Mathematicians, 1978*, pp. 245-82.

[31]   D. T. Hoang, "Searching genetic databases on Splash 2," *Proceedings IEEE Workshop on FPGAs for Custom Computing Machines (Cat. No.93TH0535-5). IEEE Comput. Soc. Press. 1993*, pp. 185-91.

[32]   D. P. Lopresti, "P-NAC: a systolic array for comparing nucleic acid sequences," *Computer*, vol. 20, pp. 98-9, 1987.

[33]   Z. Peixin, M. Martonosi, P. Ashar, and S. Malik, "Using configurable computing to accelerate Boolean satisfiability," *IEEE Transactions on Computer-Aided Design of Integrated Circuits & Systems*, vol. 18, pp. 861-8, 1999.

[34]   A. J. Elbirt and C. Paar, "An FPGA implementation and performance evaluation of the Serpent block cipher," *FPGA'00. ACM/SIGDA International Symposium on Field Programmable Gate Arrays. ACM. 2000*, pp. 33-40.

[35]   K. Hea Joung and W. H. Mangione-Smith, "Factoring large numbers with programmable hardware," *FPGA'00. ACM/SIGDA International Symposium on Field Programmable Gate Arrays. ACM. 2000*, pp. 41-8.

[36]   M. Weinhardt and W. Luk, "Pipeline vectorization," *IEEE Transactions on Computer-Aided Design of Integrated Circuits & Systems*, vol. 20, pp. 234-48, 2001.

[37]   W. J. Huang, N. Saxena, and E. J. McCluskey, "A reliable LZ data compressor on reconfigurable coprocessors," *Proceedings 2000 IEEE Symposium on Field-Programmable Custom Computing Machines (Cat. No.PR00871). IEEE Comput. Soc. 2000*, pp. 249-58.

[38]   P. Graham and B. Nelson, "Genetic algorithms in software and in hardware-a performance analysis of workstation and custom computing machine implementations," *Proceedings. IEEE Symposium on FPGAs for Custom Computing Machines (Cat. No.96TB100063). IEEE Comput. Soc. Press. 1996*, pp. 216-25.

[39]   J. Gause, P. Y. K. Cheung, and W. Luk, "Reconfigurable shape-adaptive template matching architectures," *Proceedings 10th Annual IEEE Symposium on Field-Programmable Custom Computing Machines. FCCM 2002. IEEE Comput. Soc. 2002*, pp. 98-107.

[40]   K. H. Tsoi, K. H. Lee, and P. H. W. Leong, "A massively parallel RC4 key search engine," *Proceedings 10th Annual IEEE Symposium on Field-Programmable Custom Computing Machines. FCCM 2002. IEEE Comput. Soc. 2002*, pp. 13-21.

[41]   M. Skubiszewski, "A hardware emulator for binary neural networks," *INNC 90 Paris. International Neural Network Conference. Kluwer. 1990*, pp. 555-8 vol.

[42]   M. Skubiszewski, "An exact hardware implementation of the Boltzmann machine," *Proceedings of the Fourth IEEE Symposium on Parallel and Distributed Processing (Cat. No.92TH0492-9). IEEE. 1992*, pp. 107-10.

[43]   C. Grassmann and J. K. Anlauf, "RACER - a rapid prototyping accelerator for pulsed neural networks," *Proceedings 10th Annual IEEE Symposium on Field-Programmable Custom Computing Machines. FCCM 2002. IEEE Comput. Soc. 2002*, pp. 277-8.

[44]   S. J. Melnikoff, S. F. Quigley, and M. J. Russell, "Implementing a simple continuous speech recognition system on an FPGA," *Proceedings 10th Annual IEEE Symposium on Field-Programmable Custom Computing Machines. FCCM 2002. IEEE Comput. Soc. 2002*, pp. 275-6.

[45]   H. Styles and W. Luk, "Accelerating radiosity calculations using reconfigurable platforms," *Proceedings 10th Annual IEEE Symposium on Field-Programmable Custom Computing Machines. FCCM 2002. IEEE Comput. Soc. 2002*, pp. 279-81.

[46] T. Yokota, M. Nagafuchi, Y. Mekada, T. Yoshinaga, K. Ootsu, and T. Baba, "A scalable FPGA-based custom computing machine for a medical image processing," *Proceedings 10th Annual IEEE Symposium on Field-Programmable Custom Computing Machines. FCCM 2002. IEEE Comput. Soc. 2002*, pp. 307-8.

[47] B. L. Hutchings, R. Franklin, and D. Carver, "Assisting network intrusion detection with reconfigurable hardware," *Proceedings 10th Annual IEEE Symposium on Field-Programmable Custom Computing Machines. FCCM 2002. IEEE Comput. Soc. 2002*, pp. 111-20.

[48] G. Brebner, "Single-chip gigabit mixed-version IP router on Virtex-II Pro," *Proceedings 10th Annual IEEE Symposium on Field-Programmable Custom Computing Machines. FCCM 2002. IEEE Comput. Soc. 2002*, pp. 35-44.

[49] P. Bellows, J. Flidr, T. Lehman, B. Schott, and K. D. Underwood, "GRIP: a reconfigurable architecture for host-based gigabit-rate packet processing," *Proceedings 10th Annual IEEE Symposium on Field-Programmable Custom Computing Machines. FCCM 2002. IEEE Comput. Soc. 2002*, pp. 121-30.

[50] G. Memik, S. O. Memik, and W. H. Mangione-Smith, "Design and analysis of a layer seven network processor accelerator using reconfigurable logic," *Proceedings 10th Annual IEEE Symposium on Field-Programmable Custom Computing Machines. FCCM 2002. IEEE Comput. Soc. 2002*, pp. 131-40.

[51] G. Estrin, "Reconfigurable computer origins: the UCLA fixed-plus-variable (F+V) structure computer," *IEEE Annals of the History of Computing*, vol. 24, pp. 3-9, 2002.

[52] S. Trimberger, *Field-programmable gate array technology*. Boston: Kluwer Academic Publishers, 1994.

[53] S. Brown and J. Rose, "FPGA and CPLD architectures: a tutorial," *IEEE Design & Test of Computers*, vol. 13, pp. 42-57, 1996.

[54] V. Betz, J. S. Rose, and A. Marquardt, *Architecture and CAD for deep-submicron FPGAs*. Boston: Kluwer Academic, 1999.

[55] Xilinx, "XC3000 Series Field Programmable Gate Arrays (XC3000A/L, XC3100A/L)," Version 3.1, 1998.

[56] Xilinx, "Virtex-II Pro Platform FPGAs: Introduction and Overview," 2003.

[57] S. Trimberger, *Field-programmable gate array technology*. Boston: Kluwer Academic Publishers, 1994.

[58] G. Borriello, C. Ebeling, S. A. Hauck, and S. Burns, "The Triptych FPGA architecture," *IEEE Transactions on Very Large Scale Integration (Vlsi) Systems*, vol. 3, pp. 491-501, 1995.

[59] C. Ebeling, L. McMurchie, S. A. Hauck, and S. Burns, "Placement and routing tools for the Triptych FPGA," *IEEE Transactions on Very Large Scale Integration (Vlsi) Systems*, vol. 3, pp. 473-82, 1995.

[60] C. Ebeling, G. Borriello, S. A. Hauck, D. Song, and E. A. Walkup, "TRIPTYCH: a new FPGA architecture," *FPGAs. International Workshop on Field Programmable Logic and Applications. Abingdon EE&CS Books. 1991*, pp. 75-90.

[61] S. Hauck, G. Borriello, S. Burns, and C. Ebeling, "Montage: an FPGA for synchronous and asynchronous circuits," *Field-Programmable Gate Arrays: Architectures and Tools for Rapid Prototyping. Second International Workshop on Field Programmable Logic and Applications. Springer-Verlag. 1993*, pp. 44-51.

[62] S. Hauck, L. Zhiyuan, and E. Schwabe, "Configuration compression for the Xilinx XC6200 FPGA," *Proceedings. IEEE Symposium on FPGAs for Custom Computing Machines (Cat. No.98TB100251). IEEE Comput. Soc. 1998*, pp. 138-46.

[63] Z. Li, K. Compton, and S. Hauck, "Configuration caching management techniques for reconfigurable computing," *Proceedings 2000 IEEE Symposium on Field-Programmable Custom Computing Machines (Cat. No.PR00871). IEEE Comput. Soc. 2000*, pp. 22-36.

[64] S. Hauck, L. Zhiyuan, and E. Schwabe, "Configuration compression for the Xilinx XC6200 FPGA," *IEEE Transactions on Computer-Aided Design of Integrated Circuits & Systems*, vol. 18, pp. 1107-13, 1999.

[65] L. Zhiyuan and S. Hauck, "Don't Care discovery for FPGA configuration compression," *FPGA'99. AGM/SIGDA International Symposium on Field Programmable Gate Arrays. ACM. 1999*, pp. 91-8.

[66] S. Hauck, "Configuration prefetch for single context reconfigurable coprocessors," *FPGA'98. ACM/SIGDA International Symposium on Field Programmable Gate Arrays. ACM. 1998*, pp. 65-74.

[67] A. DeHon, "Reconfigurable Architectures for General-Purpose Computing," in *Dept of EECS*. Cambridge: MIT, 1996.

[68] A. DeHon, "DPGA utilization and application," *FPGA '96. 1996 ACM Fourth International Symposium on Field-Programmable Gate Arrays. ACM. 1996*, pp. 115-21.

[69] A. DeHon, "DPGA-coupled microprocessors: commodity ICs for the early 21st Century," *Proceedings IEEE Workshop on FPGAs for Custom Computing Machines (Cat. No.94TH0611-4). IEEE Comput. Soc. Press. 1994*, pp. 31-9.

[70] M. J. Wirthlin and B. L. Hutchings, "A dynamic instruction set computer," *Proceedings IEEE Symposium on FPGAs for Custom Computing Machines (Cat. No.95TB8077). IEEE Comput. Soc. Press. 1995*, pp. 99-107.

[71] "XC6200 FPGA Family Data Sheet.," Xilinx Inc. 2003.

[72] C. W. Murphy, D. M. Harvey, and L. J. Nicholson, "Low cost TMS320C40/XC6200 based re-configurable parallel image processing architecture," *IEE Colloquium Reconfigurable Systems (Ref. No.1999/061). IEE. 1999*, pp. 9/1-5.

[73] N. McKay, T. Melham, and S. Kong Woei, "Dynamic specialisation of XC6200 FPGAs by partial evaluation," *Proceedings. IEEE Symposium on FPGAs for Custom Computing Machines (Cat. No.98TB100251). IEEE Comput. Soc. 1998*, pp. 308-9.

[74] R. Woods, D. Trainor, and J. P. Heron, "Applying an XC6200 to real-time image processing," *IEEE Design & Test of Computers*, vol. 15, pp. 30-8, 1998.

[75] G. Brebner, "Automatic identification of swappable logic units in XC6200 circuitry," *Field-programmable Logic and Applications. 7th International Workshop, FPL '97. Proceedings. Springer-Verlag. 1997*, pp. 173-82.

[76] R. Payne, "Run-time parameterised circuits for the Xilinx XC6200," *Field-programmable Logic and Applications. 7th International Workshop, FPL '97. Proceedings. Springer-Verlag. 1997*, pp. 161-72.

[77] R. Woods, S. Ludwig, J. Heron, D. Trainor, and S. Gehring, "FPGA synthesis on the XC6200 using IRIS and Trianus/Hades (or from heaven to hell and back again)," *Proceedings. The 5th Annual IEEE Symposium on Field-Programmable Custom Computing Machines (Cat. No.97TB100186). IEEE Comput. Soc. 1997*, pp. 155-64.

[78]   G. Brebner, "An interactive datasheet for the Xilinx XC6200," *Field-Programmable Logic and Applications. From FPGAs to Computing Paradigm. 8th International Workshop, FPL'98. Proceedings. Springer-Verlag. 1998*, pp. 401-5.

[79]   N. McKay and S. Singh, "Dynamic specialisation of XC6200 FPGAs by partial evaluation," *Field-Programmable Logic and Applications. From FPGAs to Computing Paradigm. 8th International Workshop, FPL'98. Proceedings. Springer-Verlag. 1998*, pp. 298-307.

[80]   S. Charlwood and P. James-Roxby, "Evaluation of the XC6200-series architecture for cryptographic applications," *Field-Programmable Logic and Applications. From FPGAs to Computing Paradigm. 8th International Workshop, FPL'98. Proceedings. Springer-Verlag. 1998*, pp. 218-27.

[81]   R. W. Hartenstein, M. Herz, and F. Gilbert, "Designing for Xilinx XC6200 FPGAs," *Field-Programmable Logic and Applications. From FPGAs to Computing Paradigm. 8th International Workshop, FPL'98. Proceedings. Springer-Verlag. 1998*, pp. 29-38.

[82]   G. Brebner, "A virtual hardware operating system for the Xilinx XC6200," *Field-Programmable Logic. Smart Applications, New Paradigms and Compilers. 6th International Workshop on Field-Programmable Logic and Applications, FPL '96 Proceedings. Springer-Verlag. 1996*, pp. 327-36.

[83]   T. Kean, B. New, and B. Slous, "A fast constant coefficient multiplier for the XC6200," *Field-Programmable Logic. Smart Applications, New Paradigms and Compilers. 6th International Workshop on Field-Programmable Logic and Applications, FPL '96 Proceedings. Springer-Verlag. 1996*, pp. 230-6.

[84]   S.-M. Ludwig, "The design of a coprocessor board using Xilinx's XC6200 FPGA-an experience report," *Field-Programmable Logic. Smart Applications, New Paradigms and Compilers. 6th International Workshop on Field-Programmable Logic and Applications, FPL '96 Proceedings. Springer-Verlag. 1996*, pp. 77-86.

[85]   S. Churcher, T. Kean, and B. Wilkie, "The XC6200 FastMap/sup TM/ processor interface," *Field-Programmable Logic and Applications. 5th International Workshop, FPL '95. Proceedings. Springer-Verlag. 1995*, pp. 36-43.

[86]   R. Razdan, K. Brace, and M. D. Smith, "PRISC software acceleration techniques," *Proceedings IEEE International Conference on Computer Design: VLSI in Computers and Processors (Cat. No.94CH35712). IEEE Comput. Soc. Press. 1994*, pp. 145-9.

[87]   Z. A. Ye, A. Moshovos, S. Hauck, and P. Banerjee, "CHIMAERA: a high-performance architecture with a tightly-coupled reconfigurable functional unit," *Proceedings of 27th International Symposium on Computer Architecture (IEEE Cat. No.RS00201). ACM. 2000*, pp. 225-35.

[88]   R. D. Wittig and P. Chow, "OneChip: an FPGA processor with reconfigurable logic," *Proceedings. IEEE Symposium on FPGAs for Custom Computing Machines (Cat. No.96TB100063). IEEE Comput. Soc. Press. 1996*, pp. 126-35.

[89]   S. A. Mahlke, D. C. Lin, W. Y. Chen, R. E. Hank, and R. A. Bringmann, "Effective compiler support for predicated execution using the hyperblock," *Sigmicro Newsletter*, vol. 23, pp. 45-54, 1992.

[90]   M. Wazlowski, L. Agarwal, T. Lee, A. Smith, E. Lam, P. Athanas, H. Silverman, and S. Ghosh, "PRISM-II compiler and architecture," *Proceedings IEEE Workshop on FPGAs for Custom Computing Machines (Cat. No.93TH0535-5). IEEE Comput. Soc. Press. 1993*, pp. 9-16.

[91]   A. K. Yeung and J. M. Rabaey, "A 2.4 GOPS data-driven reconfigurable
       multiprocessor IC for DSP," *1995 IEEE International Solid-State Circuits Conference.
       Digest of Technical Papers. ISSCC (Cat. No.95CH35753). IEEE. 1995*, pp. 108-9, 346.

[92]   A. K. W. Yeung and J. M. Rabaey, "A reconfigurable data-driven multiprocessor
       architecture for rapid prototyping of high throughput DSP algorithms," *Proceeding of the
       Twenty-Sixth Hawaii International Conference on System Sciences (Cat. No.93TH0501-
       7). IEEE. 1993*, pp. 169-78 vol.

[93]   R. A. Sutton, V. P. Srini, and J. M. Rabaey, "A multiprocessor DSP system using
       PADDI-2," *Proceedings 1998 Design and Automation Conference. 35th DAC. (Cat.
       No.98CH36175). IEEE. 1998*, pp. 62-5.

[94]   D. Genin, P. Hilfinger, J. Rabaey, C. Scheers, and H. De Man, "DSP specification
       using the Silage language," *ICASSP 90. 1990 International Conference on Acoustics,
       Speech and Signal Processing (Cat. No.90CH2847-2). IEEE. 1990*, pp. 1056-60 vol.

[95]   J. M. M. Verbakel, "The DSP description language SILAGE," *Proceedings of the Ninth
       Symposium on Information Theory in the Benelux. Werkgemeenschap Inf. &
       Communicatietheorie. 1988*, pp. 67-73.

[96]   H. Schmit, D. Whelihan, A. Tsai, M. Moe, B. Levine, and R. Reed Taylor, "PipeRench:
       A virtualized programmable datapath in 0.18 micron technology," *Proceedings of the
       IEEE 2002 Custom Integrated Circuits Conference (Cat. No.02CH37285). IEEE. 2002*,
       pp. 63-6.

[97]   C. Yuan, P. Pillai, H. Schmit, and J. P. Shen, "PipeRench implementation of the
       Instruction Path Coprocessor," *Proceedings 33rd Annual IEEE/ACM International
       Symposium on Microarchitecture. MICRO-33 2000. IEEE. 2000*, pp. 147-58.

[98]   S. C. Goldstein, H. Schmit, M. Budiu, S. Cadambi, M. Moe, and R. R. Taylor,
       "PipeRench: a reconfigurable architecture and compiler," *Computer*, vol. 33, pp. 70-7,
       2000.

[99]   R. Laufer, R. R. Taylor, and H. Schmit, "PCI-PipeRench and the SWORDAPI: a
       system for stream-based reconfigurable computing," *Seventh Annual IEEE Symposium on
       Field-Programmable Custom Computing Machines (Cat. No.PR00375). IEEE Comput.
       Soc. 1999*, pp. 200-8.

[100] S. C. Goldstein, H. Schmit, M. Moe, M. Budiu, S. Cadambi, R. R. Taylor, and R.
       Laufer, "PipeRench: a coprocessor for streaming multimedia acceleration," *ACM.
       Computer Architecture News*, vol. 27, pp. 28-39, 1999.

[101] S. C. Goldstein, H. Schmit, M. Moe, M. Budiu, S. Cadambi, R. R. Taylor, and R.
       Laufer, "PipeRench: a coprocessor for streaming multimedia acceleration," *Proceedings
       of the 26th International Symposium on Computer Architecture (Cat. No.99CB36367).
       IEEE Comput. Soc. Press. 1999*, pp. 28-39.

[102] C. Ebeling, D. C. Cronquist, and P. Franklin, "RaPiD-reconfigurable pipelined
       datapath," *Field-Programmable Logic. Smart Applications, New Paradigms and
       Compilers. 6th International Workshop on Field-Programmable Logic and Applications,
       FPL '96 Proceedings. Springer-Verlag. 1996*, pp. 126-35.

[103] C. Fisher, K. Rennie, X. Guanbin, S. G. Berg, K. Bolding, J. Naegle, D. Parshall, D.
       Portnov, A. Sulejmanpasic, and C. Ebeling, "An emulator for exploring RaPiD
       configurable computing architectures," *Field Programmable Logic and Applications. 11th
       International Conference, FPL 2001. Proceedings (Lecture Notes in Computer Science
       Vol.2147). Springer-Verlag. 2001*, pp. 17-26.

[104] D. C. Cronquist, C. Fisher, M. Figueroa, P. Franklin, and C. Ebeling, "Architecture design of reconfigurable pipelined datapaths," *Proceedings 20th Anniversary Conference on Advanced Research in VLSI. IEEE Comput. Soc. 1999*, pp. 23-40.

[105] D. C. Cronquist, P. Franklin, S. G. Berg, and C. Ebeling, "Specifying and compiling applications for RaPiD," *Proceedings. IEEE Symposium on FPGAs for Custom Computing Machines (Cat. No.98TB100251). IEEE Comput. Soc. 1998*, pp. 116-25.

[106] C. Ebeling, D. C. Cronquist, P. Franklin, J. Secosky, and S. G. Berg, "Mapping applications to the RaPiD configurable architecture," *Proceedings. The 5th Annual IEEE Symposium on Field-Programmable Custom Computing Machines (Cat. No.97TB100186). IEEE Comput. Soc. 1997*, pp. 106-15.

[107] E. Mirsky and A. DeHon, "MATRIX: a reconfigurable computing architecture with configurable instruction distribution and deployable resources," *Proceedings. IEEE Symposium on FPGAs for Custom Computing Machines (Cat. No.96TB100063). IEEE Comput. Soc. Press.*, pp. 157-66, 1996.

[108] M. B. Taylor, J. Kim, J. Miller, D. Wentzlaff, F. Ghodrat, B. Greenwald, H. Hoffman, P. Johnson, J.-W. Lee, W. Lee, A. Ma, A. Saraf, M. Seneski, N. Shnidman, V. Strumpen, M. Frank, S. Amarasinghe, and A. Agarwal, "The raw microprocessor: a computational fabric for software circuits and general-purpose programs," *IEEE Micro*, vol. 22, pp. 25-35, 2002.

[109] E. Waingold, M. Taylor, D. Srikrishna, V. Sarkar, W. Lee, V. Lee, J. Kim, M. Frank, P. Finch, R. Barua, J. Babb, S. Amarasinghe, and A. Agarwal, "Baring it all to software: Raw machines," *Computer*, vol. 30, pp. 86-93, 1997.

[110] C. R. Rupp, M. Landguth, T. Garverick, E. Gomersall, H. Holt, J. M. Arnold, and M. Gokhale, "The NAPA adaptive processing architecture," *Proceedings. IEEE Symposium on FPGAs for Custom Computing Machines (Cat. No.98TB100251). IEEE Comput. Soc. 1998*, pp. 28-37.

[111] M. B. Gokhale, J. M. Stone, and E. Gomersall, "Co-synthesis to a hybrid RISC/FPGA architecture," *Journal of VLSI Signal Processing*, vol. 24, pp. 165-80, 2000.

[112] M. B. Gokhale and J. M. Stone, "NAPA C: compiling for a hybrid RISC/FPGA architecture," *Proceedings. IEEE Symposium on FPGAs for Custom Computing Machines (Cat. No.98TB100251). IEEE Comput. Soc. 1998*, pp. 126-35.

[113] P. Bertin, D. Roncin, and J. Vuillemin, "Introduction to programmable active memories," *Systolic Array Processors. Contributions by Speakers at the International Conference on Systolic Arrays. Prentice Hall. 1989*, pp. 301-9.

[114] D. Belosloudtsev, P. Bertin, R. K. Bock, P. Boucard, V. Doersing, P. Kammel, S. Khabarov, F. Klefenz, W. Krischer, A. Kugel, L. Lundheim, R. Maenner, L. Moll, K. H. Noffz, A. Reinsch, D. Ronein, M. Shand, J. Vuillemin, and R. Zoz, "Programmable active memories in real-time tasks: implementing data-driven triggers for LHC experiments," *Nuclear Instruments & Methods in Physics Research Section A-Accelerators Spectrometers Detectors & Associated Equipment*, vol. 356, pp. 457-67, 1995.

[115] R. A. Keaney, L. H. C. Lee, D. J. Skellern, J. Vuillemin, and M. Shand, "Implementation of long constraint length Viterbi decoders using programmable active memories," *11th Australian Microelectronics Conference. Microelectronics, Meeting the Needs of Modern Technology. Proceedings MICRO '93. Inst. Radio & Electron. Eng. 1993*, pp. 52-7.

[116] P. Bertin, D. Roncin, and J. Vuillemin, "Programmable active memories: a performance assessment," *Parallel Architectures and Their Efficient Use. First Heinz Nixdorf Symposium Proceedings. Springer-Verlag. 1993*, pp. 119-30.

[117] S. Ogrenci Memik, E. Bozorgzadeh, R. Kastner, and M. Sarrafzadeh, "Strategically programmable systems," *Proceedings of the Reconfigurable Architecture Workshop*, 2001.

[118] E. Bozorgzadeh, S. Ogrenci Memik, R. Kastner, and M. Sarrafzadeh, "Strategically programmable systems," in *Computer Engineering Handbook*: CRC Press, 2001.

[119] E. Boerger, "Architecture Design and Validation Methods," Springer-Verlag, 2000.

[120] M. Zelkowitz, "Advances in Computers," vol. 56. London: Academic Press, 2002.

[121] E. A. Lee and A. Sangiovanni-Vincentelli, "A framework for comparing models of computation," *IEEE Transactions on Computer-Aided Design of Integrated Circuits and Systems*, vol. 17, pp. 1217-29, 1998.

[122] V. Sassone, M. Nielsen, and G. Winskel, "A classification of models for concurrency," *Proceedings of the International Conference on Concurrency Theory*, 1993.

[123] M. Nielsen, V. Sassone, and G. Winskel, "Relationships between models of concurrency," *Proceedings of the Decade of Concurrency. Reflections and Perspectives.*, 1994.

[124] R. D. Tennent, "The denotational semantics of programming languages," *Communications of the ACM*, vol. 19, pp. 437-53, 1976.

[125] J. Davis, C. Hylands, J. Janneck, E. A. Lee, J. Liu, X. Liu, S. Neuendorffer, S. Sachs, M. Stewart, K. Vissers, P. Whitaker, and Y. Xiong, "Overview of the Ptolemy project," University of California, Berkeley UCB/ERL M01/11, 2001.

[126] D. Harel and A. Pnueli, "On the Development of Reactive Systems," in *Logic and Models for Verification and Specification of Concurrent Systems*, vol. 13, K. R. Apt, Ed.: Springer Verlag, 1985.

[127] G. Berry, "Real Time Programming: Special Purpose or General Purpose Languages," in *Information Processing*, G. Ritter, Ed.: Elsevier Science Publishers, 1992.

[128] E. A. Lee, "Embedded Software," in *Advances in Computers*, vol. 56, M. Zelkowitz, Ed. London: Academic Press, 2002.

[129] J. E. Hopcroft, R. Motwani, and J. D. Ullman, *Introduction to automata theory, languages, and computation*, 2nd ed. Boston: Addison-Wesley, 2001.

[130] C. A. R. Hoare, "Communicating sequential processes," *Communications of the ACM*, vol. 21, pp. 666-77, 1978.

[131] G. Berry and G. Gonthier, "The ESTEREL synchronous programming language: design, semantics, implementation," *Science of Computer Programming*, vol. 19, pp. 87-152, 1992.

[132] A. Benveniste and P. Le Guernic, "Hybrid dynamical systems theory and the Signal language," *IEEE Transactions on Automatic Control*, vol. 35, pp. 535-46, 1990.

[133] P. Caspi, D. Pilaud, N. Halbwachs, and J. A. Plaice, "LUSTRE: a declarative language for programming, synchronous systems," *Proceedings of the Symposium on Principles of Programming Languages*, 1987.

[134] M. Jourdan, F. Lagnier, F. Maraninchi, and P. Raymond, "A multiparadigm language for reactive systems," *Proceedings of the International Conference on Computer Languages*, 1994.

[135] G. Kahn, "The semantics of a simple language for parallel programming," *Proceedings of the IFIP Congress*, 1974.

[136] E. A. Lee and T. M. Parks, "Dataflow process networks," *Proceedings of the IEEE*, vol. 83, pp. 773-801, 1995.

[137] E. A. Lee and D. G. Messerschmitt, "Synchronous data flow," *Proceedings of the IEEE*, vol. 75, pp. 1235-45, 1987.

[138] H. Trickey, "Flamel: A high level hardware compiler," *IEEE Transactions on Computer-Aided Design of Integrated Circuits and Systems*, vol. CAD-6, pp. 259-69, 1987.

[139] S. A. Hayati, A. C. Parker, and J. J. Granacki, "Representation of control and timing behavior with applications to interface synthesis," *Proceedings of the IEEE International Conference on Computer Design: VLSI in Computers and Processors*, 1988.

[140] A. Girault, L. Bilung, and E. A. Lee, "Hierarchical finite state machines with multiple concurrency models," *IEEE Transactions on Computer-Aided Design of Integrated Circuits & Systems*, vol. 18, pp. 742-60, 1999.

[141] D. Harel, "Statecharts: A Visual Formulation for Complex Systems," *Science of Computer Programming*, vol. 8, pp. 231-274, 1987.

[142] A. P. David Harel, Jeanette P. Schmidt, R. Sherman, "On the Formal Semantics of Statecharts," *Proceedings of the Proceedings of the Symposium on Logic in Computer Science*, Ithaca, New York, 1987.

[143] A. N. David Harel, "The STATEMATE Semantics of Statecharts," *ACM Transactions on Software Engineering and Methodology*, vol. 5, pp. 293-333, 1996.

[144] K. S. Lothar Thiele, Dirk Ziegenbein, Rolf Ernst, Jürgen Teich, "FunState - An Internal Design Representation for Codesign," *Proceedings of the International Conference on Computer Aided Design*, San Jose, CA, 1999.

[145] G. G. Gérard Berry, "The Esterel Synchronous Programming Language: Design, Semantics, Implementation," *Science of Computer Programming*, vol. 19, pp. 87-152, 1992.

[146] G. Berry, "The Foundations of Esterel," in *Proof, Language, and Interaction: Essays in Honour of Robin Milner*, C. S. a. M. T. G. Plotkin, Ed.: MIT Press, 1998.

[147] G. Berry, "A Hardware Implementation of Pure Esterel," *Proceedings of the Proceedings of the International Workshop on Formal Methods in VLSI Design*, Miami, FL, 1991.

[148] S. N. Frank Vahid, and Daniel D. Gajski, "SpecCharts: A VHDL Front-End for Embedded Systems," *IEEE Transactions On Computer-Aided Design of Integrated Circuits and Systems*, vol. 14, pp. 694-706, 1995.

[149] M. C. Eylon Caspi, Randy Huang, Joseph Yeh, John Wawrzynek, and André DeHon, "Stream Computations Organized for Reconfigurable Execution (SCORE): Extended Abstract," *Proceedings of the Conference on Field Programmable Logic and Applications (FPL)*, Villach, Austria, 2000.

[150] M. Gokhale, J. Stone, J. Arnold, and M. Kalinowski, "Stream-oriented FPGA computing in the Streams-C high level language," *Proceedings 2000 IEEE Symposium on Field-Programmable Custom Computing Machines (Cat. No.PR00871). IEEE Comput. Soc. 2000*, pp. 49-56.

[151] D. D. Gajski, J. Zhu, R. Dömer, A. Gerstlauser, and S. Zhoa, *SpecC: Specification Language and Methodology*. Boston: Kluwer Academic Publishers, 2000.

[152] C. Lukai, P. Kritzinger, M. Olivares, and D. Gajski, "Top-down system level design methodology using SpecC, VCC and SystemC," *Proceedings of the Design, Automation and Test in Europe Conference*, 2002.

[153] T. Grotker, *System Design with SystemC*. Boston: Kluwer Academic Publishers, 2002.

[154] S. Y. Liao, "Towards a new standard for system-level design," *Proceedings of the International Workshop on Hardware/Software Codesign*, 2000.

[155] H. Chao, S. Ravi, A. Raghunathan, and N. K. Jha, "High-level synthesis of distributed logic-memory architectures," *IEEE/ACM International Conference on Computer Aided Design. IEEE/ACM Digest of Technical Papers (Cat. No.02CH37391). IEEE. 2002*, pp. 564-71.

[156] P. Diniz and J. Park, "Automatic synthesis of data storage and control structures for FPGA-based computing engines," *Proceedings 2000 IEEE Symposium on Field-Programmable Custom Computing Machines (Cat. No.PR00871). IEEE Comput. Soc. 2000*, pp. 91-100.

[157] L. Semeria, K. Sato, and G. De Micheli, "Synthesis of hardware models in C with pointers and complex data structures," *IEEE Transactions on Very Large Scale Integration (VLSI) Systems*, vol. 9, pp. 743-56, 2001.

[158] W. Qin, S. Rajagopalan, M. Vachharajani, H. Wang, X. Zhu, D. August, K. Keutzer, S. Malik, and L. S. Peh, "Design tools for application specific embedded processors," *Embedded Software. Second International Conference, EMSOFT 2002. Proceedings (Lecture Notes in Computer Science Vol.2491). Springer-Verlag. 2002*, pp. 319-33.

[159] D. I. August, K. Keutzer, S. Malik, and A. R. Newton, "A disciplined approach to the development of platform architectures," *Microelectronics*, vol. 33, pp. 881-90, 2002.

[160] K. Keutzer, S. Malik, and A. R. Newton, "From ASIC to ASIP: the next design discontinuity," *Proceedings 2002 IEEE International Conference on Computer Design: VLSI in Computers and Processors. IEEE Comput. Soc. 2002*, pp. 84-90.

[161] E. Caspi, M. Chu, R. Huang, J. Yeh, J. Wawrzynek, and A. DeHon, "Stream computations organized for reconfigurable execution (SCORE)," *Field-Programmable Logic and Applications. Roadmap to Reconfigurable Computing. 10th International Conference, FPL 2000. Proceedings (Lecture Notes in Computer Science Vol.1896). Springer-Verlag. 2000*, pp. 605-14.

[162] K. Strehl, L. Thiele, M. Gries, D. Ziegenbein, R. Ernst, and J. Teich, "FunState-an internal design representation for codesign," *IEEE Transactions on Very Large Scale Integration (Vlsi) Systems*, vol. 9, pp. 524-44, 2001.

[163] D. Harel, "Statecharts: a visual formalism for complex system," *Science of Computer Programming*, vol. 8, pp. 231-74, 1987.

[164] S. A. Edwards, "ESUIF: an open Esterel compiler," *SO - Elsevier. Electronic Notes in Theoretical Computer Science, vol.65, no.5, 2002, Netherlands.*

[165] M. W. Hall, J. M. Anderson, S. P. Amarasinghe, B. R. Murphy, L. Shih-Wei, E. Bugnion, and M. S. Lam, "Maximizing multiprocessor performance with the SUIF compiler," *Computer*, vol. 29, pp. 84-9, 1996.

[166] M. D. Smith and G. Holloway, "An introduction to machine SUIF and its portable libraries for analysis and optimization," Division of Engineering and Applied Sciences, Harvard University.

[167] L. Shih-Wei, A. Diwan, R. P. Bosch, Jr., A. Ghuloum, and M. S. Lam, "SUIF Explorer: an interactive and interprocedural parallelizer," *ACM. Sigplan Notices (Acm Special Interest Group on Programming Languages)*, vol. 34, pp. 37-48, 1999.

[168] A. W. Lim, L. Shih-Wei, and M. S. Lam, "Blocking and array contraction across arbitrarily nested loops using affine partitioning," *ACM. Sigplan Notices (Acm Special Interest Group on Programming Languages)*, vol. 36, pp. 103-12, 2001.

[169] A. W. Lim, G. I. Cheong, and M. S. Lam, "An affine partitioning algorithm to maximize parallelism and minimize communication," *Conference Proceedings of the 1999 International Conference on Supercomputing. ACM. 1999*, pp. 228-37.

[170] J. T. Oplinger, D. L. Heine, and M. S. Lam, "In search of speculative thread-level parallelism," *1999 International Conference on Parallel Architectures and Compilation Techniques (Cat. No.PR00425). IEEE Comput. Soc. 1999*, pp. 303-13.

[171] K. Bondalapati, P. Diniz, P. Duncan, J. Granack, M. Hall, R. Jain, and H. Ziegler, "DEFACTO: a design environment for adaptive computing technology," *Parallel and Distributed Processing. 11th IPPS/SPDP'99 Workshops Held in Conjunction with the 13th International Parallel Processing Symposium and 10th Symposium on Parallel and Distributed Processing. Proceedings. Springer-Verlag. 1999*, pp. 570-8.

[172] W. M. W. Hwu, S. A. Mahlke, W. Y. Chen, P. P. Chang, N. J. Warter, R. A. Bringmann, R. G. Ouellette, R. E. Hank, T. Kiyohara, G. E. Haab, J. G. Holm, and D. M. Lavery, "The superblock: an effective technique for VLIW and superscalar compilation," *Journal of Supercomputing*, vol. 7, pp. 229-48, 1993.

[173] D. I. August, W. W. Hwu, and S. A. Mahlke, "A framework for balancing control flow and predication," *Proceedings. Thirtieth Annual IEEE/ACM International Symposium on Microarchitecture (Cat. No.97TB100184). IEEE Comput. Soc. 1997*, pp. 92-103.

[174] L. Carter, B. Simon, B. Calder, and J. Ferrante, "Predicated static single assignment," *Proceedings of the 1999 International Conference on Parallel Architectures and Compilation Techniques*, 1999.

[175] J. Ferrante, K. J. Ottenstein, and J. D. Warren, "The program dependence graph and its use in optimization," *ACM Transactions on Programming Languages & Systems*, vol. 9, pp. 319-49, 1987.

[176] R. Ernst, J. Henkel, and T. Benner, "Hardware-software cosynthesis for microcontrollers," *IEEE Design & Test of Computers*, vol. 10, pp. 64-75, 1993.

[177] R. K. Gupta and G. De Micheli, "Constrained software generation for hardware-software systems," *Proceedings of the Third International Workshop on Hardware/Software Codesign (Cat. No.94TH0700-5). IEEE Comput. Soc. Press. 1994*, pp. 56-63.

[178] W. Hardt and R. Camposano, "Criteria for hardware/software partitioning in system design," *GI 93. Anwenderforum. GI/ITG-Workshop. CAD-Umgebungen und Methoden des Entwurfes von Schaltkreisen und Systemen (GI CAD Environments and Design Methods for Switching Devices and Systems). Tech. Univ. Dresden. 1993*, pp. 7-13.

[179] F. Vahid and L. Thuy Din, "Extending the Kernighan/Lin heuristic for hardware and software functional partitioning," *Design Automation for Embedded Systems*, vol. 2, pp. 237-61, 1997.

[180] S. Edwards, L. Lavagno, E. A. Lee, and A. Sangiovanni-Vincentelli, "Design of embedded systems: formal models, validation, and synthesis," *Proceedings of the IEEE*, vol. 85, pp. 366-90, 1997.

[181] M. Baleani, F. Gennari, J. Yunjian, Y. Patel, R. K. Brayton, and A. Sangiovanni-Vincentelli, "HW/SW partitioning and code generation of embedded control applications on a reconfigurable architecture platform," *Proceedings of the Tenth International Symposium on Hardware/Software Codesign. CODES 2002 (IEEE Cat. No.02TH8627). ACM. 2002*, pp. 151-6.

[182] J. Harkin, T. M. McGinnity, and L. P. Maguire, "Partitioning methodology for dynamically reconfigurable embedded systems," *IEE Proceedings-E Computers & Digital Techniques*, vol. 147, pp. 391-6, 2000.

[183] L. Yanbing, T. Callahan, R. Harr, U. Kurkure, and J. Stockwood, "Hardware-software co-design of embedded reconfigurable architectures," *Proceedings 2000. Design Automation Conference. (IEEE Cat. No.00CH37106). ACM. 2000*, pp. 507-12.

[184] G. Holloway and M. D. Smith, "Machine SUIF Control Flow Graph Library," Division of Engineering and Applied Sciences, Harvard University 2002.

[185] G. Holloway and M. D. Smith, "Machine-SUIF SUIFvm Library," Division of Engineering and Applied Sciences, Harvard University 2002.

[186] G. Holloway and M. D. Smith, "Machine-SUIF Machine Library," Division of Engineering and Applied Sciences, Harvard University 2002.

[187] S. A. Mahlke, D. C. Lin, W. Y. Chen, R. E. Hank, and R. A. Bringmann, "Effective compiler support for predicated execution using the hyperblock," *Proceedings of the International Symposium on Microarchitecture*, 1992.

[188] C. Lee, M. Potkonjak, and W. H. Mangione-Smith, "MediaBench: a tool for evaluating and synthesizing multimedia and communications systems," *Proceedings of the International Symposium on Microarchitecture*, 1997.

[189] "Quick Reconfiguration via Micro-Sequencers," Technical Report 2002.

[190] M. C. Rinard and P. C. Diniz, "Commutativity analysis: a new analysis framework for parallelizing compilers," *Proceedings of the Programming Language Design and Implementation*, 1996.

[191] M. C. Rinard and P. C. Diniz, "Commutativity analysis: a new analysis technique for parallelizing compilers," *ACM Transactions on Programming Languages and Systems*, vol. 19, pp. 942-91, 1997.

[192] R. Rugina and M. Rinard, "Pointer analysis for multithreaded programs," *Proceedings of the Programming Language Design and Implementation (PLDI)*, 1999.

[193] R. Rugina and M. Rinard, "Automatic parallelization of divide and conquer algorithms," *Proceedings of the Symposium on Principles and Practice of Parallel Programming*, 1999.

[194] K. Zee and M. Rinard, "Write barrier removal by static analysis," *SIGPLAN Notices*, vol. 37, pp. 32-41, 2002.

[195] A. Salcianu and M. Rinard, "Pointer and escape analysis for multithreaded programs," *Proceedings of the Symposium on Principles and Practice of Parallel Programming*, 2001.

[196] F. Vivien and M. Rinard, "Incrementalized pointer and escape analysis," *SIGPLAN Notices*, vol. 36, pp. 35-46, 2001.

[197] R. Rugina and M. Rinard, "Symbolic bounds analysis of pointers, array indices, and accessed memory regions," *Proceedings of the Programming Language Design and Implementation (PDLI)*, 2000.

[198] J. Whaley and M. Rinard, "Compositional pointer and escape analysis for Java programs," *Proceedings of the Object-Oriented Programming, Systems, Languages and Applications (OOPSLA'99)*, 1999.

[199] S. Horwitz, T. Reps, and D. Binkley, "Interprocedural slicing using dependence graphs," *ACM Transactions on Programming Languages & Systems*, vol. 12, pp. 26-60, 1990.

[200] R. A. Ballance, A. R. Maccabe, and K. J. Ottenstein, "The program dependence web: a representation supporting control-, data-, and demand-driven interpretation of imperative languages," *Sigplan Notices (Acm Special Interest Group on Programming Languages)*, vol. 25, pp. 257-71, 1990.

[201] R. Johnson and K. Pingali, "Dependence-based program analysis," *Sigplan Notices (Acm Special Interest Group on Programming Languages)*, vol. 28, pp. 78-89, 1993.

[202] D. Weise, R. F. Crew, M. Ernst, and B. Steensgaard, "Value dependence graphs: representation without taxation," *Conference Record of POPL '94: 21st ACM SIGPLAN-SIGACT Symposium on Principles of Programming Languages. ACM. 1994*, pp. 297-310.

[203] L. Chunho, M. Potkonjak, and W. H. Mangione-Smith, "MediaBench: a tool for evaluating and synthesizing multimedia and communications systems," *Proceedings. Thirtieth Annual IEEE/ACM International Symposium on Microarchitecture (Cat. No.97TB100184). IEEE Comput. Soc. 1997*, pp. 330-5.

[204] S. A. Edwards, "An Esterel compiler for large control-dominated systems," *IEEE Transactions on Computer-Aided Design of Integrated Circuits & Systems*, vol. 21, pp. 169-83, 2002.

[205] N. Ramasubramanian, R. Subramanian, and S. Pande, "Automatic analysis of loops to exploit operator parallelism on reconfigurable systems," *Languages and Compilers for Parallel Computing. 11th International Workshop, LCPC'98. Proceedings (Lecture Notes in Computer Science Vol.1656). Springer-Verlag. 1999*, pp. 305-22.

[206] J. L. Tripp, P. A. Jackson, and B. L. Hutchings, "Sea cucumber: a synthesizing compiler for FPGAs," *Proceedings of the International Conference on Field-Programmable Logic and Applications*, 2002.

[207] M. Budiu and S. C. Goldstein, "Compiling application-specific hardware," *Field-Programmable Logic and Applications. Reconfigurable Computing Is Going Mainstream. 12th International Conference, FPL 2002. Proceedings (Lecture Notes in Computer Science Vol.2438). Springer-Verlag. 2002*, pp. 853-63.

[208] S. A. Edwards, "An Esterel compiler for large control-dominated systems," *IEEE Transactions on Computer-Aided Design of Integrated Circuits and Systems*, vol. 21, pp. 169-83, 2002.

[209] R. Rinker, M. Carter, A. Patel, M. Chawathe, C. Ross, J. Hammes, W. A. Najjar, and W. Bohm, "An automated process for compiling dataflow graphs into reconfigurable hardware," *IEEE Transactions on Very Large Scale Integration (VLSI) Systems*, vol. 9, pp. 130-9, 2001.

[210] G. De Micheli, D. Ku, F. Mailhot, and T. Truong, "The Olympus synthesis system," *IEEE Design & Test of Computers*, vol. 7, pp. 37-53, 1990.

[211] A. Jones, D. Bagchi, S. Pal, P. Banerjee, and A. Choudhary, "PACT HDL: a compiler targeting ASICs and FPGAs with power and performance optimizations," in *Power aware computing*, R. Graybill and R. Melhem, Eds. New York, NY, USA: Kluwer Academic/Plenum Publishers, 2002, pp. 169-90.

[212] M. Budiu and S. C. Goldstein, "Compiling application-specific hardware," *Proceedings of the International Conference on Field-Programmable Logic and Applications*, 2002.

[213] J. Babb, M. Rinard, C. A. Moritz, W. Lee, M. Frank, R. Barua, and S. Amarasinghe, "Parallelizing applications into silicon," *Proceedings of the Seventh Annual IEEE Symposium on Field-Programmable Custom Computing Machines*, 1999.

[214] D. Galloway, "The Transmogrifier C hardware description language and compiler for FPGAs," *Proceedings of the IEEE Symposium on FPGAs for Custom Computing Machines*, 1995.

[215] Atmel, "FPSLIC."

[216] "ARM-Based Embedded Processor PLDs," Altera Corporation 2001.

[217] Xilinx, "Virtex-II Pro Platorm FPGA Handbook," 2002.

[218] K. Bondalapati, P. Diniz, P. Duncan, J. Granack, M. Hall, R. Jain, and H. Ziegler, "DEFACTO: a design environment for adaptive computing technology," *Proceedings of the 11th IPPS/SPDP'99 Workshops Held in Conjunction with the 13th International Parallel Processing Symposium and 10th Symposium on Parallel and Distributed Processing*, 1999.

[219] M. B. Gokhale and J. M. Stone, "NAPA C: compiling for a hybrid RISC/FPGA architecture," *Proceedings of the*, 1998.

[220] I. Page, "Constructing hardware-software systems from a single description," *Journal of VLSI Signal Processing*, vol. 12, pp. 87-107, 1996.

[221] T. J. Callahan and J. Wawrzynek, "Instruction-level parallelism for reconfigurable computing," *Proceedings of the*, 1998.

[222] G. Vanmeerbeeck, P. Schaumont, S. Vernalde, M. Engels, and I. Bolsens, "Hardware/software partitioning of embedded system in OCAPI-xl," *Proceedings of the*, 2001.

[223] C. J. Alpert and A. B. Kahng, "Recent directions in netlist partitioning: a survey," *Integration, The VLSI Journal*, vol. 19, pp. 1-81, 1995.

[224] F. Vahid, G. Jie, and D. D. Gajski, "A binary-constraint search algorithm for minimizing hardware during hardware/software partitioning," *Proceedings of the*, 1994.

[225] J. Henkel, "A low power hardware/software partitioning approach for core-based embedded systems," *Proceedings of the*, 1999.

[226] A. Kalavade, "System Level Codesign of Mixed Hardware-Software Systems," UCB, PhD Dissertation ERL 95/98, September 1995.

[227] D. E. Thomas, J. K. Adams, and H. Schmit, "A model and methodology for hardware-software codesign," *IEEE Design & Test of Computers*, vol. 10, pp. 6-15, 1993.

[228] A. Kalavade and E. A. Lee, "A hardware-software codesign methodology for DSP applications," *IEEE Design & Test of Computers*, vol. 10, pp. 16-28, 1993.

[229] A. Kalavade and E. A. Lee, "The extended partitioning problem: hardware/software mapping, scheduling, and implementation-bin selection," *Design Automation for Embedded Systems*, vol. 2, pp. 125-63, 1997.

[230] R. K. Gupta and G. De Micheli, "System-level synthesis using re-programmable components," *Proceedings of the European Conference on Design Automation*, 1992.

[231] P. Zebo and K. Kuchcinski, "An algorithm for partitioning of application specific systems," *Proceedings of the European Conference on Design Automation*, 1993.

[232] R. Niemann and P. Marwedel, "Hardware/software partitioning using integer programming," *Proceedings of the European Design and Test Conference*, 1996.

[233] J. Henkel and R. Ernst, "A hardware/software partitioner using a dynamically determined granularity," *Proceedings of the Design Automation Conference*, 1997.

[234] Celoxica, "Frequently Asked Questions: DK1 Design Suite," 2001.

[235] A. Balboni, W. Fornaciari, and D. Sciuto, "Co-synthesis and co-simulation of control-dominated embedded systems," *Design Automation for Embedded Systems*, vol. 1, pp. 257-89, 1996.

[236] R. Camposano and J. Wilberg, "Embedded system design," *Design Automation for Embedded Systems*, vol. 1, pp. 5-50, 1996.

[237] M. Chiodo, D. Engels, P. Giusto, H. Hsieh, A. Jurecska, L. Lavagno, K. Suzuki, and A. Sangiovanni-Vincentelli, "A case study in computer-aided co-design of embedded controllers," *Design Automation for Embedded Systems*, vol. 1, pp. 51-67, 1996.

[238] L. Yanbing, T. Callahan, R. Harr, U. Kurkure, and J. Stockwood, "Hardware-software co-design of embedded reconfigurable architectures," *Proceedings of the Design Automation Conference*, 2000.

[239] T. Callahan and J. Wawrzynek, "Adapting Software Pipelining for Reconfigurable Computing," *Proceedings of the International Conference on Compilers, Architecture, and Synthesis for Embedded Systems (CASES)*, 2000.

[240] M. Weinhardt and W. Luk, "Pipeline vectorization," *IEEE Transactions on Computer-Aided Design of Integrated Circuits and Systems*, vol. 20, pp. 234-48, 2001.

[241] G. Wang, W. Gong, and R. Kastner, "Task Level Partitioning Using Ant System Algorithm," Department of Electrical and Computer Engineering, University of California, Santa Barbara, Technical Report 2003.

[242] M. Dorigo, V. Maniezzo, and A. Colorni, "Ant system: optimization by a colony of cooperating agents," *IEEE Transactions on Systems, Man, & Cybernetics, Part B: Cybernetics*, vol. 26, pp. 29-41, 1996.

[243] J. M. Pasteels, J. L. Deneubourg, and Fondation les treilles, *From individual to collective behavior in social insects : les Treilles Workshop*. Basel ; Boston: Birkhäuser, 1987.

[244] T. Wiangtong, P. Y. K. Cheung, and W. Luk, "Comparing three heuristic search methods for functional partitioning in hardware-software codesign," *Design Automation for Embedded Systems*, vol. 6, pp. 425-49, 2002.

[245] R. Schreiber, S. Aditya, B. R. Rau, V. Kathail, S. Mahlke, S. Abraham, and G. Snider, "High-level Synthesis of Nonprogrammable Hardware Accelerators," *Proceedings of the International Conference on Application-Specific Systems, Architectures, and Processors*, 2000.

[246] R. E. Gonzalez, "Xtensa: A Configurable and Extensible Processor," in *IEEE Micro*, vol. 20, 2000, pp. 60-70.

[247] P. M. Athanas and A. L. Abbott, "Real-time image processing on a custom computing platform," in *IEEE Computer*, vol. 28, pp. 16-25.

[248] M. Gokhale, W. Holmes, A. Kopser, S. Lucas, R. Minnich, D. Sweely, and D. Lopresti, "Building and Using A Highly Parallel Programmable Logic Array," in *IEEE Computer*, vol. 24, pp. 81-89.

[249] Z. Peixin, M. Martonosi, P. Ashar, and S. Malik, "Using Configurable Computing To Accelerate Boolean Satisfiability," *IEEE Transactions on Computer-Aided Design of Integrated Circuits and Systems*, vol. 18, pp. 861-868.

[250] R. P. S. Sidhu, A. Mei, and V. K. Prasanna, "String Matching on Multicontext FPGAs Using Self-reconfiguration," *Proceedings of the International Symposium on Field Programmable Gate Arrays*, 1999.

[251] S. C. Goldstein, H. Schmit, M. Budiu, S. Cadambi, M. Moe, and R. R. Taylor, "PipeRench: A Reconfigurable Architecture and Compiler," in *IEEE Computer*, vol. 33, 2000, pp. 70-77.

[252] C. Ebeling, D. C. Cronquist, and P. Franklin, "RaPiD-reconfigurable pipelined datapath," *Proceedings of the Proceedings of the Workshop on Field-Programmable Logic and Applications*, 1996.

[253] M. B. Taylor, J. Kim, J. Miller, D. Wentzlaff, F. Ghodrat, B. Greenwald, H. Hoffman, P. Johnson, J.-W. Lee, W. Lee, A. Ma, A. Saraf, M. Seneski, N. Shnidman, V. Strumpen, M. Frank, S. Amarasinghe, and A. Agarwal, "The raw microprocessor: a computational fabric for software circuits and general-purpose programs," in *IEEE Micro*, vol. 22, 2002, pp. 25-35.

[254] T. Callahan, J. Hauser, and J. Wawrzynek, "The Garp Architecture and C Compiler," in *IEEE Computer*, 2000.

[255] S. O. Memik, E. Bozorgzadeh, R. Kastner, and M. Sarrafzadeh, "SPS: A Strategically Programmable System," *Proceedings of the Reconfigurable Architecture Workshop*, 2001.

[256] A. DeHon, "DPGA Utilization and Application," *Proceedings of the International Symposium on Field Programmable Gate Arrays*, 1996.

[257] S. Hauck, T. W. Fry, M. M. Hosler, and J. P. Kao, "The Chimaera Reconfigurable Functional Unit," *Proceedings of the Proceedings of the Symposium on Field-Programmable Custom Computing Machines*, 1997.

[258] S. A. Edwards, "An Esterel Compiler For Large Control-Dominated Systems," *IEEE Transactions on Computer-Aided Design of Integrated Circuits and Systems*, vol. 21, pp. 169-183.

[259] A. Girault, L. Bilung, and E. A. Lee, "Hierarchical Finite State Machines With Multiple Concurrency Models," *IEEE Transactions on Computer-Aided Design of Integrated Circuits and Systems*, vol. 18, pp. 742-760.

[260] E. Bozorgzadeh, S. O. Memik, R. Kastner, and M. Sarrafzadeh, "Pattern Selection: Customized Block Allocation for Domain-Specific Programmable Systems," *Proceedings of the International Conference on Engineering of Reconfigurable Systems and Algorithms (ERSA'02)*, 2002.

[261] S. O. Memik, E. Bozorgzadeh, R. Kastner, and M. Sarrafzadeh, "A Super-Scheduler For Embedded Reconfigurable Systems," *Proceedings of the Proceedings of the International Conference on Computer-Aided Design*, 2001.

[262] R. Kastner, S. O. Memik, E. Bozorgzadeh, and M. Sarrafzadeh, "Instruction Generation for Hybrid Reconfigurable Systems," *Proceedings of the Proceedings of the International Conference on Computer-Aided Design*, 2001.

[263] P. Brisk, A. Kaplan, R. Kastner, and M. Sarrafzadeh, "Instruction Generation and Regularity Extraction for Reconfigurable Processors," *Proceedings of the Proceedings of the International Conference on Compilers, Architecture, and Synthesis for Embedded Systems*, 2002.

[264] M. R. Garey and D. S. Johnson, *Computers and Intractibility: A Guide to the Theory of NP-Completeness*.

[265] S. Cadambi and S. C. Goldstein, "CPR: A Configuration Profiling Tool," *Proceedings of the FCCM*, 1999.

[266] D. S. Rao and F. J. Kurdahi, "On Clustering For Maximal Regularity Extraction," *IEEE Transactions on Computer-Aided Design of Integrated Circuits and Systems*, vol. 12, pp. 1198-1208, 1993.

[267] P. Foggia, C. Sansone, and M. Vento, "A Performance Comparison of Five Algorithms for Graph Isomorphism," *Proceedings of the Proc. of the 3rd IAPR-TC-15 International Workshop on graph based Representation*, Italy, 2001.

[268] J. R. Ullmann, "An Algorithm for Subgraph Isomorphism," *Journal of the ACM*, vol. 23, pp. 31-42, 1976.

[269] B. D. McKay, "Practical Graph Isomorphism," *Congressus Numerantium*, vol. 30, pp. 45-87, 1981.

[270] D. C. Schmidt and L. E. Druffel, "A Fast Backtracking Algorithm to Test Directed Graphs For Isomorphism Using Distance Matrices," *Journal of the ACM*, vol. 23, pp. 433-455, 1976.

[271] L. P. Cordella, P. Foggia, C. Sansone, and M. Vento, "Evaluation Performance of the VF Graph Matching Algorithm," *Proceedings of the Proc. Of the 10th International Conference on Image Analysis and Processing*, 1999.

[272] L. P. Cordella, P. Foggia, C. Sansone, and M. Vento, "An Improved Algorithm for Matching Large Graphs," *Proceedings of the The 3rd IAPR-TC15 Workshop on Graph-based Representations*, 2001.

[273] M. W. Hall, J. M. Anderson, S. P. Amarasinghe, B. R. Murphy, L. Shih-Wei, E. Bugnion, and M. S. Lam, "Maximizing Multiprocessor Performance With The SUIF Compiler," in *IEEE Computer*, vol. 29, pp. 84-89.

[274] E. Oja and J. Karhunen, "Recursive Construction of Karhunen-Loeve Expansions for Pattern Recognition Purposes," *Proceedings of the Proceedings of the 5th International Conference on Pattern Recognition*, Miami Beach, FL, 1980.

[275] N. F. P. National Bureau of Standards, "Guidelines for Implementing and Using the NBS Data Encryption Standard," U.S. Department of Commerce April 1981 1981.

[276] J. Daemen and V. Rijmen, "The Block Cipher Rijndael," in *Smart Card Research and Applications*, J.-J. Q. a. B. Schneier, Ed.: Springer-Verlag, 2000, pp. 288-296.

[277] L. Tai, D. Knapp, R. Miller, and D. MacMillen, "Scheduling Using Behavioral Templates," *Proceedings of the Proceedings of the Design Automation Conference*, 1995.

[278] T. J. Callahan, P. Chong, A. DeHon, and J. Wawrzynek, "Fast Module Mapping and Placement for Datapaths in FPGAs," *Proceedings of the Proceedings of the International Symposium on Field Programmable Gate Arrays*, 1998.

[279] A. Chowdhary, S. Kale, P. Saripella, N. Sehgal, and R. Gupta, "A General Approach for Regularity Extraction in Datapath Circuits," *Proceedings of the International Conference on Computer-Aided Design*, 1998.

[280] R. Mehra and J. Rabaey, "Exploiting Regularity for Low-Power Design," *Proceedings of the Proceedings of the International Conference on Computer-Aided Design*, 1996.

[281] M. Kahrs, "Matching a Parts Library in a Silicon Compiler," *Proceedings of the Proceedings of the International Conference on Computer-Aided Design*, 1986.

[282] K. Keutzer, "DAGON: Technology Binding and Local Optimization by DAG Matching," *Proceedings of the Design Automation Conference*, Miami Beach, FL, 1987.

[283] S. Note, W. Geurts, F. Catthoor, and H. D. Man, "Cathedral-III: Architecture-Driven High-Level Synthesis for High Throughput DSP Applications," *Proceedings of the Proceedings of the Design Automation Conference*, 1991.

[284] M. R. Corazao, M. A. Khalaf, L. M. Guerra, M. Potkonjak, and J. M. Rabaey, "Performance Optimization Using Template Mapping for Datapath-Intensive High-Level Synthesis," *IEEE Transactions on Computer-Aided Design of Integrated Circuits and Systems*, vol. 15, 1996.

[285] K. Compton and S. Hauck, "Totem: Custom Reconfigurable Array Generation," *Proceedings of the Proceedings of the Symposium on FPGAs for Custom Computing Machines Conference*, 2001.

[286] K. Compton, A. Sharma, S. Phillips, and S. Hauck, "Flexible Routing Architecture Generation for Domain-Specific Reconfigurable Subsystems," *Proceedings of the International Symposium on Field Programmable Logic and Applications*, 2002.

[287] R. Kastner, S. Ogrenci Memik, E. Bozorgzadeh, and M. Sarrafzadeh, "Instruction generation for hybrid reconfigurable systems," *Proceedings of the International Conference on Computer Aided Design*, 2001.

[288] R. Cytron, J. Ferrante, B. K. Rosen, M. N. Wegman, and F. K. Zadeick, "An efficient method of computing static single assignment form," *Proceedings of the Sixteenth Annual ACM Symposium on Principles of Programming Languages*, 1989.

[289] P. Briggs, K. D. Cooper, T. J. Harvey, and L. T. Simpson, "Practical improvements to the construction and destruction of static single assignment form," *Software - Practice and Experience*, vol. 28, pp. 859-81, 1998.

[290] R. Cytron, J. Ferrante, B. K. Rosen, M. N. Wegman, and F. K. Zadeck, "Efficiently computing static single assignment form and the control dependence graph," *ACM Transactions on Programming Languages and Systems*, vol. 13, pp. 451-90, 1991.

[291] S. L. Graham and M. Wegman, "A fast and usually linear algorithm for global flow analysis," *Journal of the Association for Computing Machinery*, vol. 23, pp. 172-202, 1976.

[292] K. Kennedy, "A survey of data flow analysis techniques," in *Program flow analysis. Theory and applications*, S. S. Muchnick and N. D. Jones, Eds. Englewood Cliffs, NJ, USA: Prentice-Hall, 1981, pp. 5-54.

[293] J. B. Kam and J. D. Ullman, "Global data flow analysis and iterative algorithms," *Journal of the Association for Computing Machinery*, vol. 23, pp. 158-71, 1976.

[294] S. S. Muchnick, *Advanced compiler design and implementation*. San Francisco, Calif.: Morgan Kaufmann Publishers, 1997.

[295] P. Briggs, T. Harvey, and L. Simpson, "Static Single Assignment Construction," 196.

[296] A. Ranjan, K. Bazargan, S. Ogrenci, and M. Sarrafzadeh, "Fast floorplanning for effective prediction and construction," *IEEE Transactions on Very Large Scale Integration (Vlsi) Systems*, vol. 9, pp. 341-51, 2001.

[297] P. Briggs and K. D. Cooper, "Effective partial redundancy elimination," *Proceedings of the ACM SIGPLAN '94 Conference on Programming Language Design and Implementation*, 1994.

[298] P. Briggs, K. D. Cooper, and L. T. Simpson, "Value numbering," *Software - Practice and Experience*, vol. 27, pp. 701-24, 1997.

[299] B. Alpern, M. N. Wegman, and F. K. Zadeck, "Detecting equality of variables in programs," *Proceedings of the Fifteenth Annual ACM Symposium on Principles of Programming Languages*, 1988.

[300] W. Amme, N. Dalton, J. von Ronne, and M. Franz, "SafeTSA: a type safe and referentially secure mobile-code representation based on static single assignment form," *SIGPLAN Notices*, vol. 36, pp. 137-47, 2001.

[301] H. C. Torng and S. Vassiliadis, *Instruction-level parallel processors*. Los Alamitos, Calif.: IEEE Computer Society Press, 1995.

[302] R. E. Hank, S. A. Mahlke, R. A. Bringmann, J. C. Gyllenhaal, and W. W. Hwu, "Superblock formation using static program analysis," *Proceedings of the 26th Annual International Symposium on Microarchitecture (Cat. No.93TH0602-3). IEEE Comput. Soc. Press. 1993*, pp. 247-55.

[303] M. Wazlowski, L. Agarwal, T. Lee, A. Smith, E. Lam, P. Athanas, H. Silverman, and S. Ghosh, "PRISM-II compiler and architecture," *Proceedings of the IEEE Workshop on FPGAs for Custom Computing Machines*, 1993.

[304] "Open SystemC Initiative."

Index

Contents